W9-BUJ-076

The Institute of Biology's
Studies in Biology no. 37

Photosynthesis

by *D. O. Hall* Ph.D.

Reader in Biology, King's College, University of London
and

K. K. Rao Ph.D.

Honorary Lecturer, King's College, University of London

Edward Arnold

First published 1972
by Edward Arnold (Publishers) Limited,
25 Hill Street,
London, W1X 8LL

Boards edition ISBN: 0 7131 2378 8
Paper edition ISBN: 0 7131 2379 6

Printed in Great Britain by
William Clowes & Sons, Limited,
London, Beccles and Colchester

General Preface to the Series

It is no longer possible for one textbook to cover the whole field of Biology and to remain sufficiently up to date. At the same time students at school, and indeed those in their first year at universities, must be contemporary in their biological outlook and know where the most important developments are taking place.

The Biological Education Committee, set up jointly by the Royal Society and the Institute of Biology, is sponsoring, therefore, the production of a series of booklets dealing with limited biological topics in which recent progress has been most rapid and important.

A feature of the series is that the booklets indicate as clearly as possible the methods that have been employed in elucidating the problems with which they deal. Wherever appropriate there are suggestions for practical work for the student. To ensure that each booklet is kept up to date, comments and questions about the contents may be sent to the author or the Institute.

1972   INSTITUTE OF BIOLOGY
                                          41 Queen's Gate
                                          London, S.W.7

Preface

There are many ways to study and teach photosynthesis, probably due to the varied background of those scientists who contribute to the field—all the way from field ecologists, physiologists and biochemists to physical chemists and mathematicians. This is one of the reasons why the study of photosynthesis is such an exciting field since it can stimulate activity from so many different people with their varied approaches to the problem. We have not opted for any rigid approach in our presentation but tried to give the student a diverse insight into the practical and theoretical aspects of photosynthesis showing the presently accepted tenets (with our bias, of course) and the experimental basis on which our present hypotheses are based.

A reading list is provided to help students answer the inevitable questions which will arise. It can be a soul-pleasing task to solve one's own questions but we hope not too many unanswerable questions will arise which will be beyond the scope of the references cited. We are always open to questioning, and possible problem solving if any student is too baffled.

Please approach this text to learn the basis of photosynthesis and then ask the relevant questions as they arise in individual chapters.

We thank Mrs. Dorothy Dallas, Faculty of Education, King's College, for helpful comments upon the manuscript.

London, 1972   D.O.H. and K.K.R.

Contents

1.1 Ultimate energy source

The term photosynthesis literally means building up or assembly by light. As used commonly, photosynthesis is the process by which plants synthesize organic compounds from inorganic raw materials in the presence of sunlight. All forms of life in this universe require energy for growth and maintenance. Algae, higher plants and certain types of bacteria capture this energy directly from the solar radiation and utilize the energy for the synthesis of essential food materials. Animals cannot use sunlight directly as a source of energy; they obtain the energy by eating plants or by eating other animals which have eaten plants. Thus the ultimate source of all metabolic energy in our planet is the sun and photosynthesis is essential for maintaining all forms of life on earth.

We use coal, natural gas, petroleum, etc. as fuels. All these fuels are decomposition products of land and marine animals and the energy stored in these materials was captured from the solar radiation millions of years ago. Solar radiation is also responsible for the formation of wind and rain and hence the energy from windmills and hydro-electric power stations could also be traced back to the sun.

The major chemical pathway in photosynthesis is the conversion of carbon dioxide and water to carbohydrates and oxygen. The reaction can be represented by the equation,

$$CO_2 + H_2O \xrightarrow[\text{Plants}]{\text{Sunlight}} \underset{\text{Carbohydrate}}{[CH_2O]} + O_2$$

The carbohydrates formed possess more energy than the starting materials, viz. CO_2 and H_2O. By the input of the sun's energy the energy-poor compounds, CO_2 and H_2O, are converted to the energy-rich compounds, carbohydrate and O_2. The energy levels of the various reactions which lead up to the above overall equation can be expressed on an oxidation-reduction scale ('redox potential' given in volts) which tells us the energy available in any given reaction—this will be discussed later in Chapter 4. Photosynthesis can thus be regarded as a process of converting radiant energy of the sun to chemical energy of plant tissues.

1.2 The carbon dioxide cycle

The CO_2 content of the atmosphere remains almost constant in spite of its depletion during photosynthesis. All plants and animals carry out the process of respiration (in mitochondria) whereby oxygen is taken from the atmosphere by living tissues to convert carbohydrates and other tissue constituents eventually to carbon dioxide and water, with the simultaneous liberation of energy. The energy is stored up as ATP and is utilized for the normal functions of the organism. Respiration thus causes a decrease in the organic matter and oxygen content and an increase in the CO_2 content of the planet. Respiration by living organisms and combustion of carbonaceous fuels consume on an average about 10 000 tonnes of O_2 every second on the surface of the earth. At this rate all the oxygen of the atmosphere would have been used up in about 3000 years. Fortunately for us the loss of organic matter and atmospheric oxygen during respiration is counterbalanced by the production of carbohydrates and oxygen during photo-

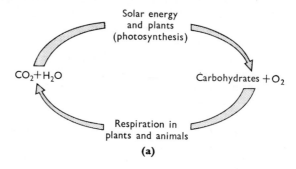

(a)

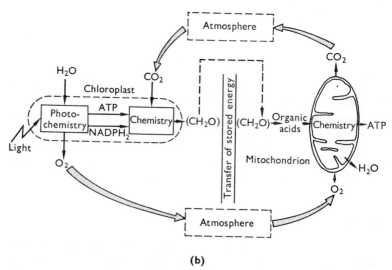

(b)

Fig. 1-1   The CO_2 and O_2 cycle in the atmosphere and the cell. (1-1(b) courtesy Professor F. R. Whatley.)

synthesis. Under ideal conditions the rate of photosynthesis in the green parts of plants is about 30 times as much as the rate of respiration in the same tissues. Thus photosynthesis is very important in regulating the O_2 and CO_2 content of the earth's atmosphere. The cycle of operations can be represented as shown in Fig. 1–1(a) and (b).

It should be made clear that the energy liberated during respiration is finally dissipated from the living organism as heat and is not available for recycling. Thus from millions of years past energy is constantly removed from the sun and wasted as heat in the earth's atmosphere. But there is still enough energy in the sun's atmosphere for photosynthesis to continue for millions of years to come.

1.3 Efficiency and turnover

The solar energy striking the earth's atmosphere every year is roughly equivalent to 520×10^{22} joules of heat. Of this about 60% is reflected back by the clouds and by the gases in the upper atmosphere. Of the remaining 40% of radiation that reaches the earth's surface only one half is in the spectral region of light that could bring about photosynthesis, the other half being weak infrared radiation (see below). Thus the annual influx of energy of photosynthetically active radiation, i.e. from violet to red light, to the earth's surface is equivalent to 100×10^{22} joules. However, some 40% of this is reflected by ocean surface, deserts, etc. and only the rest is absorbed by the plant life on land and sea. The total annual amount of biomass produced by autotrophic plants is estimated to be about 100×10^9 tonnes of carbon which is equivalent to about 170×10^{19} joules of energy. A major proportion of this organic matter is synthesized by phytoplanktons, minute plants living near the surface of the oceans. The annual food intake by the earth's human population (assuming the population to be 4000 million) is approximately 800 million tonnes or 13×10^{18} joules. Thus the average coefficient of utilization of the incident photosynthetically active radiation by the entire flora of the earth is only about 0.2% ($170 \times 10^{19}/ 100 \times 10^{22}$) and of this less than 1% ($13 \times 10^{18}/170 \times 10^{19}$) is consumed as nutrient energy by mankind.

1.4 Spectra

Light is a form of electromagnetic radiation. All electromagnetic radiation has wave characteristics and travels at the same speed of 3×10^8 m s^{-1} (c, the speed of light). But the radiations differ in wavelength, the distance between two successive peaks of the wave. Gamma rays and x-rays have very small wavelengths (less than 1000 millionth of a centimetre) while radio waves are in the order of 10^4 cm. Wave lengths of visible light are conveniently expressed by a unit called nanometer. One nanometer is 1000 millionth of a meter ($1 \text{ nm} = 10^{-9}$ m). It has been known since the time of Isaac Newton that white light can be separated into a spectrum, resembling the rainbow, by passing light through a prism. The visible portion of this spectrum ranges from the violet at about 380 nm to the red at 750 nm.

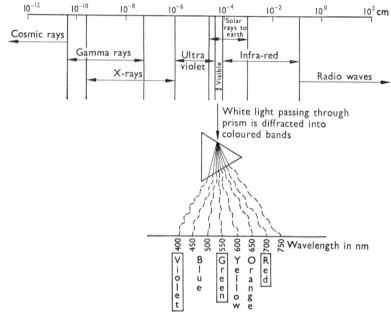

Fig. 1–2   Spectra of electromagnetic radiation.

The atmosphere of the sun consists mainly of hydrogen. The energy of the sun is derived from the fusion of four hydrogen nuclei to form a helium nucleus. $4\,^1_1\text{H} \rightarrow \,^4_2\text{He} + 2e^- + h\nu$ (energy). The energy liberated during the nuclear fission maintains the surface temperature of the sun around 6000 °C. The sun radiates energy representing the entire electromagnetic spectrum but the earth's atmosphere is transparent only to part of the infrared and ultraviolet light and all the visible light. The ultraviolet waves which are somewhat shorter than the shortest visible light waves are absorbed by the oxygen and ozone of the upper atmosphere. This is fortunate since ultraviolet radiations are harmful to living organisms. At 6000 °C, the temperature of the sun, the maximum intensity of emitted light lies in the orange part of the visible spectrum, around 600 nm.

1.5   Quantum theory

In 1900 Max Planck enunciated the theory that the transfer of radiation energy within a hot object involved discrete 'units' of energy called quanta.

Planck's quantum theory can be expressed mathematically as $E = h\nu$ where E is the energy of a single quantum of radiation, ν is the frequency of the radiation (frequency is the number of waves transmitted in unit time), and h a constant. The Planck's constant (h) has the dimensions of the product of energy and time and its value in the c.g.s. system is 6.625×10^{-34} J s. Planck's theory proposes that an oscillator of fundamental frequency (ν) would take up energy $h\nu$, $2h\nu$, $3h\nu \rightarrow nh\nu$, but it could not acquire less than a whole number of energy quanta. Five years later Albert Einstein extended Planck's theory to light and proposed that light energy is transmitted not in a continuous stream but only in individual units or quanta. The energy of a single quantum of light or *photon* is the product of the frequency of light and Planck's constant, i.e. $E = h\nu$. Since frequency is inversely related to wavelength, it follows that photons of short wave light are more energetic than photons of light of longer wavelength, i.e. at one end of the spectrum, photons of blue light are more energetic than those of red light at the other end.

For photosynthesis to take place the pigments present in plant tissues should absorb the energy of a photon at a characteristic wavelength and then utilize this energy to initiate a chain of photosynthetic chemical events. We will learn later that an electron is ejected from the pigment immediately after the absorption of a suitable quantum of light. It should be emphasized that a photon cannot transfer its energy to two or more electrons nor can the energy of two or more photons combine to eject an electron. Thus the photon should possess a critical energy to excite a single electron from the pigment molecule and initiate photosynthesis. This accounts for the low efficiency of infrared radiation in plant photosynthesis since there is insufficient energy in the quantum of infrared light. Certain bacteria, however, contain pigments which absorb infrared radiation and carry out photosynthesis which is quite different from plant-type photosynthesis in that no O_2 is evolved during the process (see Chapter 7).

1.6   Energy units

According to Einstein's law of photochemical equivalence a single molecule will react only after it has absorbed one photon of energy ($h\nu$). Hence one mole (gram-molecule) of a compound must absorb N ($N = 6.024 \times 10^{23}$, the Avogadro number) photons of energy, i.e. $Nh\nu$, to start a reaction. The total energy of photons absorbed by one mole of a compound is called an Einstein.

Let us calculate the energy of a mole (or Einstein, i.e. 6.024×10^{23} quanta) of red light of wavelength 650 nm (6.5×10^{-7} m). The frequency, $\nu = c/\lambda$ = speed of light/wavelength of light

$$\nu = 3 \times 10^8 / 6.5 \times 10^{-7} = 4.6 \times 10^{14}$$

$E = Nh\nu$, i.e. Energy = number of molecules × Planck's constant × frequency

$$\therefore \; E = 6.024 \times 10^{23} \times 6.625 \times 10^{-34} \times 4.6 \times 10^{14}$$

$$= 18.37 \times 10^4 \text{ joules} = \text{energy of one Einstein of red light}$$

$$\text{or } E = 1.837 \times 10^5 / 4.184 \times 10^3 = 43.91 \text{ kcal}$$

(One kilocalorie, kcal, is equal to 4.184×10^3 joules.) Thus 1 mole of red light at 650 nm contains 18.37×10^4 joules of energy.

The energy of photons can also be expressed in terms of electron volts. An electron volt, eV, is the energy required by an electron when it falls through a potential of 1 volt, which is equal to 1.6×10^{-19} joules. If 1 mole of a substance acquires an average energy of 1 eV the total energy of the mole (6.024×10^{23} molecules) can be calculated to be 9.64×10^4 joules. Thus the energy of 1 mole of 650 nm light is equal to 1.9 eV ($18.37 \times 10^4 / 9.64 \times 10^4$).

Table 1   Energy levels of visible light

Wavelength	Colour	Joules per mole	kcal per mole	Electron volts per mole
700 nm	Red	17.10×10^4	40.86	1.77
650 nm	Orange-red	18.37×10^4	43.91	1.91
600 nm	Yellow	19.94×10^4	47.67	2.07
500 nm	Blue	23.93×10^4	57.20	2.48
400 nm	Violet	29.92×10^4	71.50	3.10

History and Progress of Ideas

2.1 Early discoveries

In the early half of the seventeenth century the Flemish physician van Helmont grew a willow tree in a bucket of soil feeding the soil with rain water only. He observed that in five years' time the tree grew to a considerable size though the amount of soil in the bucket did not diminish significantly. van Helmont naturally concluded that the material of the tree came from the *water* used to wet the soil. In 1727 the English botanist Stephen Hales published a book in which he observed that plants used mainly *air* as the nutrient during their growth. Between 1771 and 1777 the famous English chemist Joseph Priestley (who was one of the discoverers of oxygen) conducted a series of experiments on combustion and respiration and came to the conclusion that green plants were able to reverse the respiratory processes of animals. Priestley burnt a candle in an enclosed volume of air and showed that the resultant air could no longer support burning. A mouse kept in the residual air died. A green sprig of mint, however, continued to live in the residual air for weeks. At the end of this time Priestley found that a candle could burn in the reactivated air and a mouse could breathe in it. We now know that the burning candle used up the *oxygen* of the enclosed air which was replenished by the photosynthesis of the green mint. A few years later the Dutch physician, Jan Ingenhousz, discovered that plants evolved oxygen only in *sunlight* and also that only the *green* parts of the plant carried out this process.

Jean Senebier, a Swiss minister, confirmed the findings of Ingenhousz and observed further that plants used as nourishment *carbon dioxide* 'dissolved in water'. Early in the nineteenth century another Swiss scholar de Saussure, studied the quantitative relationships between the CO_2 taken up by a plant and the amount of organic matter and O_2 produced and came to the conclusion that *water* was also consumed by plants during assimilation of CO_2. In 1817 two French chemists, Pelletier and Caventou, isolated the green substance in leaves and named it *chlorophyll*. Another milestone in the history of photosynthesis was the enunciation in 1845 by Robert Mayer, a German physician, that plants transform energy of sunlight into chemical *energy*. By the middle of the last century the phenomenon of photosynthesis could be represented by the relationship

$$CO_2 + H_2O + \text{light} \xrightarrow[\text{plant}]{\text{green}} O_2 + \text{organic matter} + \text{chemical energy}$$

Accurate determinations of the ratio of CO_2 consumed to O_2 evolved during photosynthesis were carried out by the French plant physiologist Boussingault. He found in 1864 that the photosynthetic ratio—the volume of O_2 evolved to the volume of CO_2 used up—is almost unity. In the same year the German botanist Sachs (who also discovered plant respiration) demonstrated the formation of *starch* grains during photosynthesis. Sachs kept some green leaves in the dark for some hours to deplete them of their starch content. He then exposed one half of a starch-depleted leaf to light and left the other half in the dark. After some time the whole leaf was exposed to iodine vapour. The illuminated portion of the leaf turned dark violet due to the formation of starch-iodine complex; the other half did not show any colour change.

The direct connection between oxygen evolution and chloroplasts of green leaves and also the correspondence between the action spectrum of photosynthesis and the absorption spectrum of chlorophyll (see Chapter 4) were demonstrated by Engelmann in 1880. He placed a filament of the

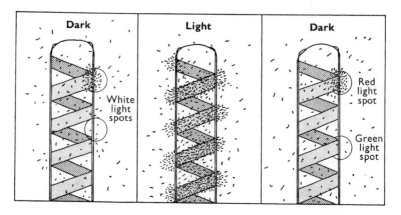

Fig. 2-1 Engelmann's experiment for studying photosynthesis using the alga *Spirogyra* and motile bacteria. The alga has a spiral chloroplast and the bacteria migrate towards regions of higher O_2 concentration. Left: Illumination with a spot of white light. Centre: Complete illumination with white light. Right: Illumination with spots of red and green light. Note the lack of O_2 evolution in green light.

green alga *Spirogyra*, with its spirally arranged chloroplasts, on a microscope slide together with a suspension of an oxygen-requiring, motile bacteria. The slide was kept in a closed chamber in the absence of air and illuminated. Motile bacteria would move towards regions of greater O_2 concentration. After a period of illumination the slide was examined under a microscope and the bacterial population counted. Engelmann found that the bacteria were concentrated around the green bands of the algal filament. In another series of experiments he illuminated the alga with a spectrum of light by interposing a prism between the light source and the microscope stage. The largest number of bacteria surrounded those parts of the algal filament that were in the blue and red regions of the spectrum. The chlorophylls present in the alga absorbed blue and red light; since light has to be absorbed to bring about photosynthesis Engelmann concluded that chloro-phylls are the active photoreceptive *pigments* for photosynthesis. The state of knowledge on photosynthesis at the beginning of this century could be represented by the equation:

$$(CO_2)_n + H_2O + light \xrightarrow[\text{plant}]{\text{green}} (O_2)_n + starch + chemical\ energy$$

2.2 Further work related to techniques

Though by the beginning of this century the overall reaction of photo-synthesis was known, the discipline of biochemistry had not advanced enough to understand the mechanism of reduction of carbon dioxide to carbohydrates. It should be admitted that even now we know very little about certain aspects of photosynthesis. Attempts were made early to study the effects of light intensity, temperature, carbon dioxide concentration, etc. on the overall yields of photosynthesis. Though plants of divergent species were used in these studies, most of the determinations were carried out with unicellular green algae, *Chlorella* and *Scenedesmus*, and the uni-cellular flagellate *Euglena*. Unicellular plants are more suitable for quantita-tive research since they can be grown in all laboratories under fairly standard conditions. They can be suspended uniformly in aqueous buffer solutions and aliquots of the suspension can be transferred with a pipette as though they were true solutions. Chloroplasts are best prepared and studied from leaves of higher plants, the most common being spinach leaves as they are usually available fresh in the market and can be grown quite easily; peas and lettuce are also sometimes used.

Since CO_2 is fairly soluble and O_2 is relatively insoluble in water, during photosynthesis in a closed system there will be a change in gas pressure. The Warburg respirometer (adapted by Otto Warburg in 1920) is exten-sively used in studies involving the action of light on photosynthetic systems by measuring the changes in the O_2 volume of the system (see 'MANO-METRIC TECHNIQUES', UMBREIT, BURRIS and STAUFFER, Burgess Publ. Co., U.S.A., for details).

The oxygen electrode is another convenient instrument to measure up-take or liberation of O_2 during a reaction. The electrode works on the prin-ciple of polarography and is sensitive enough to detect O_2 concentrations of the order of 10^{-8} moles/cm^3 (0.01 millimolar). The apparatus consists of platinum wire sealed in plastic as cathode, and an anode of circular silver wire bathed in a saturated KCl solution. The electrodes are separated from the reaction mixture by an O_2 gas-permeable teflon membrane. The re-action mixture in the plastic (or glass) container is stirred constantly with a small magnetic stirring rod. When a voltage is applied across the two elec-trodes, with the platinum electrode negative to the reference electrode, the oxygen in the solution undergoes electrolytic reduction. The flow of current in the system between 0.5 and 0.8 V varies in a linear relationship to the partial pressure of the oxygen in solution. The instrument is usually operated at a voltage of about 0.6 V. The current liberated is measured by connecting the electrode set up to a suitable recorder. The whole apparatus is kept at a constant temperature by circulating water from a controlled

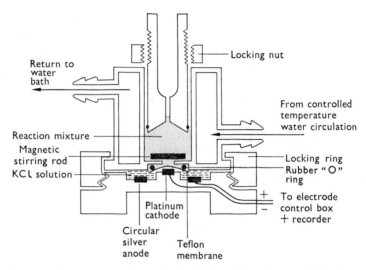

Fig. 2–2 The oxygen electrode (Rank Bros., Bottisham, Cambridge).

temperature water source. The effects of light and of various chemicals on photosynthesis are measured using the oxygen electrode. The O_2 electrode has the advantage over the Warburg method in that rapid and continuous measurements of O_2 evolution can be made. However, the Warburg apparatus can measure up to 20 reaction mixtures simultaneously while the O_2 electrode measures reactions one at a time.

2.3 Limiting factors

The extent of photosynthesis performed by a plant depends on a number of internal and external factors. The chief internal factors are the structure of the leaf and its chlorophyll content, the accumulation within the chloroplasts of the products of photosynthesis, the influence of protoplasmic enzymes and the presence of minute amounts of mineral constituents. The external factors are the quality and quantity of light incident on the leaves, the ambient temperature and the concentration of carbon dioxide and oxygen in the surrounding atmosphere.

2.3.1 *The effect of light intensity*

The effect of light intensity on the photosynthetic activity of a healthy suspension of *Chlorella* cells is illustrated in Fig. 2–3. At low light intensities the rate of photosynthesis, as measured by oxygen evolution, increases linearly in proportion to light intensity. This region of the curve, marked X, is known as the light-limiting region. With more and more light intensity, photosynthesis becomes less efficient until after about 10 000 lux

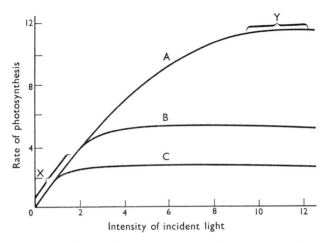

Fig. 2–3 Effect of external factors on rate of photosynthesis. A, Effect of light intensity at 25 °C and 0.4% CO_2; B, at 15 °C and 0.4% CO_2; C, at 25 °C and 0.01% CO_2. All units in the graph are arbitrary.

(1000 foot candles) increasing light intensity produces no further effect on the rate of photosynthesis. This is indicated by the horizontal parts of the curves in the figure. This plateau region designated Y is the light saturation region. If the rate of photosynthesis is to be raised in this region, factors other than light intensity would have to be adjusted. The amount of sunlight falling on a clear day in many places on the earth is about 10 000 footcandles (1 foot-candle is a measure of the luminous intensity at a distance of 1 foot from a standard candle, 1 f.c. = 10 lux). Thus, except for plants growing in thick forests and in shade, there is often sufficient sunlight incident on the plants to saturate their photosynthetic capacity. The energy of the extreme blue (400 nm) and red (800 nm) light quanta differs only by a factor of two and the photons in this wavelength range are qualitatively efficient to start photosynthesis though, as we shall later see, the leaf pigments preferentially absorb light of certain definite wavelengths.

2.3.2   Effect of temperature

A comparison of the curves A and B in the figure shows that at low light intensities the rate of photosynthesis is the same at 15 °C and at 25 °C. The reactions in the light-limiting region, like true photochemical reactions, are not sensitive to temperature. At higher light intensities, however, the rate of photosynthesis is much higher at 25 °C than at 15 °C. So, factors other than mere photon absorption influence photosynthesis in the light-saturation region. Most temperate climate plants function well between 10 °C and 35 °C, the optimum temperature being around 25 °C.

2.3.3 *Effect of CO₂ concentration*

In the light-limiting region the rate of photosynthesis is not affected by lowering the CO_2 concentration, as shown by curve C in Fig. 2–3. Thus, it can be inferred that CO_2 does not participate directly in the photochemical reaction. But at light intensities above the light-limiting region, photosynthesis is appreciably enhanced by increasing the CO_2 concentration. Photosynthesis by some crop plants, in short-term experiments, increased linearly with increasing CO_2 concentration up to about 0.5% though continued exposure to this high CO_2 concentration injured the leaves. Very good rates of photosynthesis can be obtained with a CO_2 content of about 0.1%. The average CO_2 content of the atmosphere is about 0.03 to 0.04%. Therefore plants in their normal environment do not have enough CO_2 to make maximum use of sunlight falling on them.

2.4 Light and dark reactions; flashing light experiments

As early as 1905 the British plant physiologist F. F. Blackman interpreted the shape of the light saturation curves by suggesting that photosynthesis is a two-step mechanism involving a photochemical or light reaction and a non-photochemical or dark reaction. The dark reaction which is enzymic is slower than the light reaction and hence at high light intensities the rate of photosynthesis is entirely dependent upon the rate of the dark reaction. The light reaction has a low or zero temperature coefficient while the dark reaction has a high temperature coefficient, characteristic of enzymic reactions. It should be clearly understood that the so-called dark reaction can proceed both in light and in darkness.

The light and dark reactions can be separated by using flash illuminations lasting fractions of a second. Light flashes lasting less than a millisecond (10^{-3} s) can be produced either mechanically by placing a slit in a rotating disc in the path of a steady light beam, or electrically by loading up a condenser and discharging it through a vacuum tube. Nowadays ruby lasers emitting red light at 694 nm are used as radiation source. In 1932 Emerson and Arnold illuminated suspensions of *Chlorella* cells with condenser flashes lasting about 10^{-5} s. They measured the rate of oxygen evolution in relation to the energy of the flashes, the duration of the dark intervals between the flashes, and the temperature of the cell suspension. Flash saturation occurred in normal cells when only one out of 2500 chlorophyll molecules had received a flash quantum. Emerson and Arnold concluded that the maximum yield of photosynthesis is not determined by the number of chlorophyll molecules capturing the light but by the number of enzyme molecules which carry out the dark reaction. They also observed that for dark time intervals (between successive flashes) greater than 0.06 s the yield of oxygen per flash was independent of the dark time interval; for shorter dark periods the yield increased progressively. Thus the dark reaction determining the saturation rate of photosynthesis takes about 0.06 s for completion. The average dark reaction time was calculated to be about 0.02 s. Emerson and Arnold calculated also the quantum yield (see Chapter 4) of photosynthesis from flashing light experiments.

2.5   Recent discoveries and formulations

The state of knowledge in the field of photosynthesis at the turn of this century could have been represented by the equation

$$CO_2 + H_2O \xrightarrow[\text{chlorophyll}]{\text{light}} (CH_2O) + O_2 \quad (\varDelta G = 48 \times 10^4 \text{ joules } (114 \text{ kcal}))$$

Prior to about 1930 many investigators in the field believed that the primary reaction in photosynthesis was splitting of carbon dioxide by light to carbon and oxygen; the carbon was subsequently reduced to carbohydrates by water in a different series of reactions. Two important discoveries in the 1930s changed this viewpoint. By this time a variety of bacterial cells (see Chapter 7) were found to assimilate CO_2 and synthesize carbohydrates without the use of light energy. The Dutch microbiologist C. B. van Neil did comparative studies of plant and bacterial photosynthesis and showed that some bacteria can assimilate CO_2 in light without evolving O_2. Such bacteria would not grow photosynthetically unless they were supplied with a suitable hydrogen donor substrate. Photosynthesis could be represented, according to van Neil, by the general equation

$$CO_2 + 2H_2A \xrightarrow[\text{chlorophyll}]{\text{light}} (CH_2O) + H_2O + 2A$$

where H_2A is the oxidizable substrate. van Neil suggested that photosynthesis of green plants and algae is a special case in which H_2A is H_2O and $2A$ is O_2. The primary photochemical act in plant photosynthesis would be the splitting of water to yield an oxidant (OH) and a reductant (H). The primary reductant (H) could then bring about the reduction of CO_2 to cell materials and the primary oxidant (OH) could be eliminated through a reaction to liberate O_2 and reform H_2O. The overall equation of photosynthesis for green plants, after van Neil, is

$$CO_2 + 4H_2O \xrightarrow[\text{chlorophyll}]{\text{light}} (CH_2O) + 3H_2O + O_2$$

which is a sum of three individual steps:

(i) $4H_2O \xrightarrow[\text{green pigments}]{\text{light}} 4(OH) + 4H$

(ii) $4H + CO_2 \longrightarrow (CH_2O) + H_2O$

(iii) $4(OH) \longrightarrow 2H_2O + O_2$

The reaction sequences clearly show that the oxygen is evolved from water and not from CO_2.

The second important observation was made in 1937 by R. Hill of Cambridge University. Hill separated the photosynthesizing particles (chloroplasts) of green leaves from the respiratory particles by differential centrifugation of a homogenate of leaf tissues. Hill's chloroplasts did not evolve O_2 when illuminated as such (due to possible damage of the chloroplasts during isolation) but did so when suitable electron acceptors (oxidants) like potassium ferrioxalate or potassium ferricyanide were added to the illuminated suspension. One molecule of O_2 was evolved for every four equivalents of oxidant reduced photochemically. Later many quinones and dyes were found to be reduced by illuminated chloroplasts. The chloroplasts, however, failed to reduce CO_2, the natural electron acceptor of photosynthesis. This phenomenon, now known as the Hill reaction, is a light-driven transfer of electrons from water to non-physiological oxidants (Hill reagents) against the chemical potential gradient. The significance of the Hill reaction lies in the demonstration of the fact that photochemical O_2 evolution can be separated from CO_2 reduction in photosynthesis.

The decomposition of water, and the resulting liberation of O_2 during photosynthesis, was established by Ruben and Kamen in California in 1941. They exposed photosynthesizing cells to water enriched in oxygen isotope of mass 18 (^{18}O). The isotopic composition of the oxygen evolved was the same as that of the water and not that of CO_2 used. Kamen and Ruben also discovered the radioactive isotope ^{14}C which was successfully used by Bassham, Benson and Calvin in California to trace the path of carbon in photosynthesis (Chapter 6). Calvin and co-workers showed that reduction of CO_2 to sugars proceeded by dark enzymic reactions and also that two molecules of reduced pyridine mucleotide [$NADPH_2$] and three molecules of ATP were required for the reduction of every molecule of CO_2. The role of pyridine nucleotides and ATP in the respiration of tissues was already established by this time. The photosynthetic reduction of NADP to $NADPH_2$ by isolated chloroplasts with simultaneous evolution of O_2 was demonstrated in 1951 by three different laboratories. Arnon, Allen and Whatley in 1954 also demonstrated cell-free photosynthesis, i.e. the assimilation of CO_2 and evolution of O_2 by isolated spinach chloroplasts. Proteins like ferredoxin, plastocyanin, ferredoxin reductase and cytochromes b and f, which participate in the transfer of electrons in photosynthesis, were isolated from chloroplasts within a decade.

To conclude, healthy green leaves on illumination generate $NADPH_2$ and ATP. The reducing power of $NADPH_2$ and the energy of hydrolysis of ATP are utilized in the dark enzymic reactions to reduce CO_2 to carbohydrates.

The photosynthetic apparatus is that part of the leaf cell which contains the ingredients for absorbing light and for channelling the energy of the excited pigment molecules into a series of chemical and enzymatic re-actions. Engelmann's experiments (Chapter 2) have shown that chloro-phylls are the pigments responsible for capturing light quanta. Knowledge concerning the actual cellular structure in which chlorophyll is located come from light and electron microscopy, and from cell fractionation techniques. In green algae and in higher plants the chlorophyll is contained in a cellular plastid called the *chloroplast*. Electron microscopic pictures show that chloroplasts in higher plants, e.g. spinach, tobacco, are saucer-shaped bodies 4 to 10 μm in diameter and 1 μm in thickness (1μm $= 10^{-6}$m) with an outer membrane or envelope separating it from the rest of the cytoplasm (Plates 1 and 2; Fig. 3–1). The number of chloroplasts per

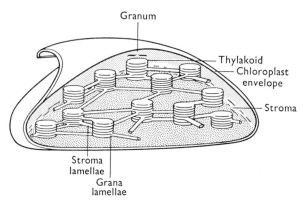

Fig. 3–1 Cut-away representation of a chloroplast to show three-dimen-sional structure.

cell, in higher plants, varies from one to more than a hundred depending upon the particular plant and on the growth conditions. In many plants the chloroplasts are able to reproduce themselves by a simple division.

Internally the chloroplast is comprised of a system of lamellae or flattened *thylakoids* which are arranged in stacks in certain dense green regions known as *grana* (Plate 3 and Fig. 3–2). Each lamella in the chloroplast may contain two double-layer membranes. The grana are embedded in a colourless matrix called the *stroma* and the whole chloroplast is surrounded by a bounding double membrane, the *chloroplast envelope*. Within a chloroplast the grana are interconnected by a system of loosely arranged membranes called the stroma lamellae or 'frets.' The detailed structure of the thylakoids is shown in Fig. 3–3—two interpretations are shown since it is impossible to decide at present as to which is correct. They probably both contain ele-

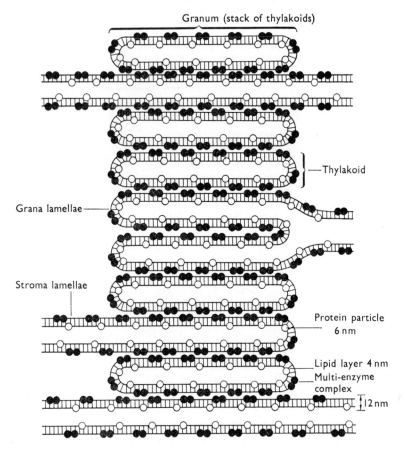

Fig. 3–2 Interpretation of the structure of a chloroplast granum from freeze-etching electron microscopy. The membranes are composed of a central lipid layer covered on both sides by attached or partially embedded protein or lipoprotein particles. (Redrawn from MÜHLETHALER, MOOR and SZOR-KOWSKI (1965) *Planta*, **67**, 305.)

ments of the true three-dimensional structure. MÜHLETHALER interprets his newer freeze etching techniques as showing that the middle layer of the membrane is mainly composed of globular proteins which are covered on both sides with a lipid monolayer; a lipid bilayer structure is only present between the globular proteins. This important concept is being actively investigated.

The lamellar structure is found not only in chloroplasts of higher plants but also in algal chloroplasts. In algae the shapes of the chloroplast are varied—Plates 4 and 5 show chloroplasts from a red and a green alga. The

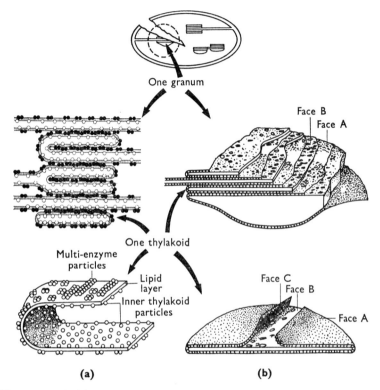

Fig. 3–3   Thylakoid structure based on interpretations of freeze etching (a) After MÜHLETHALER (Fig. 3–2); (b) After BRANTON and PARK (1967), *J. Ultrastruct. Res.* **191**, 283. Redrawn from BRANTON (1968), *Photophysiology: Current Topics* (ed. A. C. GIESE). Academic Press, New York, pp. 197–224.

most primitive algae, the blue-greens, do not contain chloroplasts as such. The photosynthetic material present in these organisms consists of parallel layers of lamellar membranes traversing through the cytoplasm.

It is possible to fractionate the chloroplasts of higher plants so that the green lamellae are separated from the colourless stroma matrix. The lamellar membranes in which the chlorophyll is embedded are approximately half lipid and half proteins in chemical composition. The proteins catalyze the enzyme reactions and give mechanical strength to the membranes while the presence of lipids facilitates energy storage and offers selective permeability of sugars, salts, substrates, etc. The quantum conversion of light energy and associated electron transport reactions of photosynthesis occur in the lamellae. The stroma contains many soluble proteins including the enzymes of the Benson–Calvin cycle (Chapter 6) which carry out the dark phase reduction of CO_2 to carbohydrates.

3.1   Isolation of chloroplasts from leaves

The first photosynthetically active chloroplasts were isolated by Hill; these preparations were active only in O_2 evolution coupled to the reduction of non-physiological electron acceptors (see Chapter 2). Arnon and Whatley isolated chloroplasts in isotonic sodium chloride, i.e. about 0.35 M or 2%. Their preparations were capable of photoreduction of NADP and photophosphorylation but were able to fix CO_2 only at low rates, although they contained all the enzymes of the Benson–Calvin CO_2 fixation cycle. These chloroplasts appeared intact under the light microscope but electron microscope pictures of the preparation indicated that they had lost their outer membranes and were naked lamellar systems. Such preparations are called 'Class II' chloroplasts (Plate 6). Recently Walker has developed techniques for the isolation of 'Class I' (complete) chloroplasts which retain their outer envelopes and which can fix CO_2 at rates between 30% and 75% of those of whole leaves. See HALL, *Nature*, **235**, 125 (1972) for discussion of chloroplast types.

Two methods used in our laboratory for the preparation of chloroplasts are given below. All solutions and apparatus used should be pre-cooled in ice. The preparation should be carried out as quickly as possible. See also Chapter 5.

Preparation 1   Procedure of Whatley and Arnon (modified)
Grinding medium:

0.35 M NaCl; 0.04 M tris/HCl buffer pH 8.0

Cut 25 g of spinach leaves into small pieces 0.5–1 cm long. Place in an M.S.E. Atomix (or domestic blender) with 50 cm³ grinding medium. Blend for 10 s on low and 20 s on high speed. Filter the homogenate through a nylon bag (or 4 layers of cheesecloth) into a centrifuge tube. Centrifuge 4 min at 2000 g. Discard the supernatant. Resuspend the pellet in 2 cm³ 0.35 M NaCl using a small piece of absorbent cotton wool wrapped round the end of a glass rod. This preparation consists of Class II chloroplasts. Broken chloroplasts are prepared by diluting the suspension with 10 vol of water to give a NaCl concentration of 0.035 M.

Preparation 2   Procedure of Walker.
Grinding medium:

Sorbitol	0.33 M
$MgCl_2$	0.005 M
$Na_4P_2O_7 \cdot 10H_2O$	0.01 M

Adjust pH of above mixture to 6.5 with HCl and add sodium isoascorbate to a final concentration of 0.002 M just before use.

Resuspending medium:

Sorbitol   0.33 M

$MgCl_2$ 0.001 M

$MnCl_2$ 0.001 M

EDTA   0.002 M

HEPES (N-2-hydroxyethyl piperazine-n-2-ethanesulphonicacid) 0.05 M.

Adjust pH to 7.6 with NaOH.

Homogenize 50 g of chilled spinach leaves with 200 cm^3 of freshly-made grinding medium for 3 to 5 s in a domestic blender. Squeeze the macerate through 2 layers of cheesecloth and filter through 8 layers of cheesecloth into 50 cm^3 plastic centrifuge tubes. Centrifuge rapidly at 0 °C from rest to 4000 g to rest in approximately 90 s. Resuspend the pellet gently using a glass rod and a small piece of absorbent cotton in 1 cm^3 of resuspending medium. This procedure should produce a suspension of chloroplasts, 50%–80%, Class I (complete), which would be capable of high rates of CO_2 fixation.

3.2   Chloroplast pigments

All photosynthetic organisms contain one or more organic pigments capable of absorbing visible radiation which will initiate the photochemical reactions of photosynthesis. These pigments can be extracted from most

Table 2   The pigments

Type of pigment	Characteristic absorption maxima (nm) (in organic solvents)	Occurrence
Chlorophylls		
Chlorophyll *a*	420, 660	All higher plants and algae
Chlorophyll *b*	435, 643	All higher plants and green algae
Chlorophyll *c*	445, 625	Diatoms and brown algae
Chlorophyll *d*	450, 690	Red algae
Carotenoids		
α-Carotene	420, 440, 490	Higher plants and some algae
β-Carotene	425, 450, 480	Some plants
Luteol	425, 445, 475	Green algae, red algae and higher plants
Violaxanthol	425, 450, 475	Higher plants
Fucoxanthol	425, 450, 475	Diatoms and brown algae
Phycobilins		
Phycoerythrins	490, 546, 576	Red algae and in some blue-green algae
Phycocyanins	618	Blue-green algae and in some red algae

leaves by boiling them in alcohol or in other organic solvents. From the alcoholic extract individual pigments can be separated by chromatography on a column of powdered sugar, as was shown by the Russian botanist Tswett in 1906. The three major classes of pigments found in plants and algae are the chlorophylls, the carotenoids and the phycobilins. The chlorophylls and carotenoids are insoluble in water but the phycobilins are soluble in water. The carotenoids and phycobilins are called the accessory photosynthetic pigments since the quanta absorbed by these pigments can be transferred to chlorophyll. Table 2 gives the absorption characteristics of these pigments. The photosynthetic pigments of bacteria are discussed in Chapter 7.

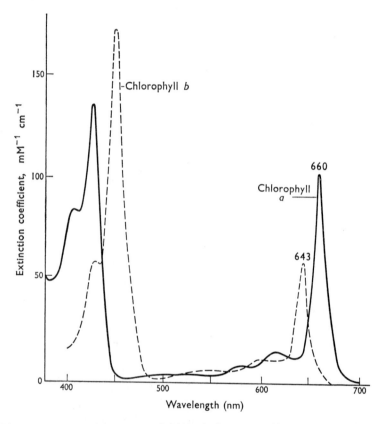

Fig. 3-4   Absorption spectra of chlorophylls extracted in acetone. (Redrawn from ZSCHEILE and COMAR (1941), *Bot. Gaz.*, **102**, 463.)

Chlorophylls are the pigments that give plants their characteristic green colour. They are insoluble in water but soluble in organic solvents. Chlorophyll *a* is bluish-green and chlorophyll *b* is yellowish-green. Chlorophyll *a* is present in all photosynthetic organisms which evolve O_2. Chlorophyll b is present (about one third of the content of chlorophyll a) in leaves of higher plants and in green algae. The absorption maxima of chlorophyll a and chlorophyll b in ether are respectively at 660 and 643 nm as shown in Fig. 3–5; in acetone the peaks are at 663 and 645 nm. However, careful spectroscopic investigations of the living cell indicate the presence of more than one form of chlorophyll a in vivo (a broad peak between 670 and 680 nm). These forms of chlorophyll a may be associated in different ways with the lamellae and probably may have different photochemical functions.

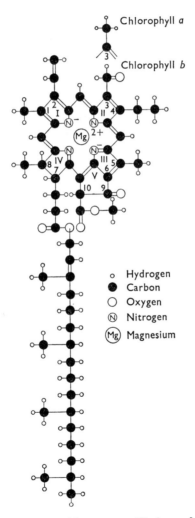

Fig. 3–5   Chlorophyll a and b structure. (Redrawn from HEATH, 1969.)

The molecular formula for chlorophyll a is $C_{55}H_{72}N_4O_5Mg$ and for chlorophyll b is $C_{55}H_{70}N_4O_6Mg$. The structural formula of chlorophyll was determined by Fischer in Germany in 1940 from degradative studies; the structure was confirmed by the complete synthesis of the molecule by Woodward at Harvard in 1960. The chlorophyll molecule (Fig. 3–5) contains a porphyrin 'head' and a phytol 'tail'. The polar (water soluble) porphyrin nucleus is made up of a tetrapyrrole ring and a magnesium atom. In the cell, electron microscopists think that the chlorophyll is sandwiched between protein and lipid layers of the chloroplasts lamellae. The porphyrin part of the molecule is bound to the protein while the phytol chain extends into the lipid layer since it is soluble in lipids.

The optical absorption curves of chlorophyll a and chlorophyll b intersect at 652 nm. A solution of chlorophyll at a concentration of 1 mg per cm^3 has an optical density of 34.5 at 652 nm. Arnon has devised the following method for the determination of the chlorophyll content of a chloroplast suspension by measuring its absorption at 652 nm. Dilute 0.1 cm^3 of chloroplast suspension with 20 cm^3 of 80% acetone, mix, and filter through Whatman No. 1 filter paper. Read the absorbancy of the solution at 652 nm in a colorimeter or spectrophotometer in a 1 cm light path cell against 80% acetone as reference. Multiply the absorbance by 5.8 to give mg of chlorophyll per cm^3 of the original chloroplast suspension.

The carotenoids are yellow or orange pigments found in all photosynthesizing cells. Their colour in the leaves is normally masked by chlorophyll, but in the autumn season when chlorophyll disintegrates the yellow pigments become visible. Carotenoids contain a conjugated double bond system of the polyene type. They are usually either hydrocarbons (carotenes) or oxygenated hydrocarbons (carotenols or xanthophylls) of 40 carbon chains built up from isoprene subunits (Fig. 3–6). They have triple-banded absorption spectra in the region from about 400 to 550 nm. The carotenoids are situated in the chloroplast lamellae in close proximity to the chlorophyll. The energy absorbed by the carotenoids may be transferred to chlorophyll a for photosynthesis. In addition the carotenoids may protect the chlorophyll molecules from too much photo-oxidation in excessive light.

Red marine algae and the primitive blue-green algae contain a group of pigments known as phycobilins. They are of two kinds: the red phycoerythrins found in red algae and the blue phycocyanins present in blue-green algae. Phycobilins are structurally related to the porphyrins of chlorophyll a but they do not have the phytyl side chain, nor do they contain magnesium. They are soluble in water. The phycoerythrins absorb light in the middle of the visible spectrum. This enables the red algae living under the sea to perform photosynthesis in the dim bluish-green light reaching the lower surfaces of the ocean—blue and red light is absorbed by the surface layers of seawater. The deeper under the sea a red alga lives the more phycoerythrin it contains in relation to chlorophyll. Phycocyanin is found in the blue-green algae which live on the surface layers of lakes and on land. The energy absorbed by the phycobilins is transferred to the chlorophylls for photo-

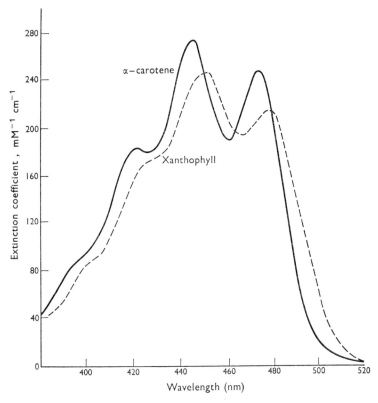

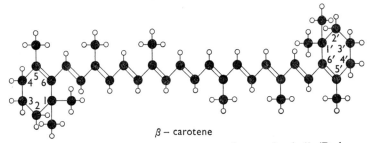

β – carotene

Fig. 3-6 Absorption spectra of α-carotene and a xanthophyll. (Redrawn from ZSCHEILE *et al.* (1942), *Plant Physiol.*, **17**, 331.) Structure of β-carotene. (Redrawn from HEATH, 1969.)

chemical processes. Thus higher plants and algae, during the course of evolution, have developed various pigments to capture the available solar radiation most efficiently and to carry out photosynthesis. The relative

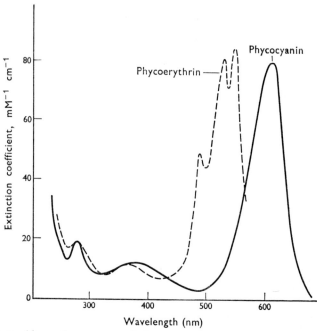

Fig. 3–7 Absorption spectra of phycoerythrin from red algae and phycocyanin from blue-green algae. (Redrawn from SVEDBURG and KATSURAI (1929), *J. Am. Chem. Soc.*, **51**, 3573.)

abundance of these pigments depends upon the species, the location of the plant, the seasons, etc.

In addition to the pigments the lamellae are composed of many proteins, lipids, quinones and metal ions. The role of some of these constituents in photosynthesis has been established by the technique of difference absorption spectroscopy (see Chapter 4). Two cytochromes, cytochrome b_6 and cytochrome f, found in chloroplasts are involved in photosynthetic electron transport. The blue copper-containing protein plastocyanin, the non-heme iron protein ferredoxin and the flavoprotein ferredoxin-NADP reductase are also located in the chloroplasts. Plastoquinone is believed to be involved in the earlier stages of electron transfer from excited chlorphyll molecules. Zinc, iron, magnesium and manganese are some of the metal ions found in chloroplast lamellae.

3·3   The photosynthetic unit

Functionally chlorophyll molecules act in groups. A photosynthetic unit is conceived as a group of pigments and other molecules utilizing the transfer of excitation energy as a mechanism by which the reaction centre communicates with an antenna of light-harvesting pigments as shown in Fig. 3–8. According to this concept a single quantum of energy absorbed anywhere in a set of about 250 chlorophyll molecules migrates to a reaction centre and promotes an electron transfer event. The following observations lead to the idea of the existence of a photosynthetic unit.

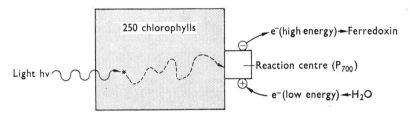

Fig. 3–8 Diagrammatic representation of the photosynthetic unit. One photon of light reacts in a unit of 250 chlorophyll molecules containing one P_{700} reaction centre.

1. Approximately 8 quanta of light absorbed by chlorophyll are required for the photosynthetic reduction of 1 CO_2 molecule and the evolution of 1 O_2 molecule. If each chlorophyll molecule in a plant can react photochemically a sufficiently intense flash of light should bring about the evolution of 1 O_2 for every 8 chlorophyll molecules present. However, the flashing light experiments of Emerson and Arnold (Chapter 2) on *Chlorella* suspensions showed that the maximum yield per flash was 1 O_2 molecule for about 2500 chlorophyll molecules, that is a single quantum of light is absorbed by about 300 chlorophyll molecules.

2. Gaffron and Wohl calculated that a single chlorophyll molecule in a dimly illuminated plant will absorb a light quantum only once in several minutes. At this rate a single molecule of chlorophyll will require nearly an hour to capture the light quanta needed for the evolution of one molecule of O_2. But when a plant is illuminated the maximum rates of CO_2 uptake and O_2 evolution are quickly established. So Gaffron and Wohl postulated that the energy harvested by a large set of chlorophyll molecules is conducted to a single reaction centre.

3. Gaffron and co-workers also observed that certain golden-yellow leaves of tobacco, with very little chlorophyll, may reach a nearly normal rate of photosynthesis at very high light intensities. These leaves would have contained a larger proportion of the special type of chlorophyll molecule which is in direct touch with the components of the electron transfer chain.

4. Difference spectroscopy has revealed the role of certain special components like P_{700} (by Kok in 1963) and cytochrome (by Duysens in 1961), in the photochemical electron transfer reactions. There is one molecule of light-reacting cytochrome and one P_{700} for every 250 chlorophyll molecules in higher plants and algae.

The lamellar membrane obtained by fractionation of isolated spinach chloroplasts can be further disrupted by sonic vibration into fragments consisting of 5 or 6 subunits; the fragments are capable of photosynthetic electron transport and phosphorylation. The subunits in the fragments have been called 'quantasomes' by Park and collaborators and can be seen as regular repeating arrays (Plate 7) under the electron microscope after suitable staining. The quantasome is an oblate sphere of approximate dimensions of $18 \times 16 \times 10$ nm and an overall molecular weight of two million. Each quantasome contains about 230 chlorophyll molecules, 2 cytochromes, 10 non-heme irons, 2 manganese and 2 copper ions, along with much protein and lipid material. Park and his associates suggested that these quantasomes were the morphological entity corresponding to the photosynthetic unit. However, there are doubts as to the exact role of the quantasomes and as to whether the quantasomes really exist as such in leaves; they could be artefacts generated during the isolation and fixation of chloroplasts for electron microscopy but the concept of a quantasome is still appealing in equating it with the photosynthetic unit.

Light absorption and emission by atoms and molecules

The most stable states of atoms are those in which the valence electrons are distributed, in accordance with the Pauli principle, into the quantum states of least energy, i.e. the electrons are in their ground states of energy level. When light is absorbed by an atom in the ground state the whole energy of the quantum ($h\nu$) is added to it, and the electrons are lifted to an energy-rich excited state. This is illustrated in Fig. 4–1, taking the helium atom as an

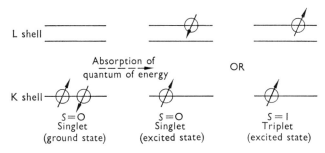

Fig. 4–1   Energy levels of electrons in the helium atom.

example. The time needed for the entire process is of the order of 10^{-15} s. If an atom has an even number of electrons, the spins of these electrons are usually arranged in opposite directions and cancel each other out so that the total electronic spin of the atom s=o (singlet state). For an atom with an odd number of electrons the net spin is s=$\frac{1}{2}$ (doublet state). If an atom has an even number of electrons and if the electronic spins are in the parallel direction the net spin s=1 and the atom is said to be in the triplet state. These states are illustrated in Fig. 4–1. The transition from the ground to the excited state of an atom by the absorption of a single quantum of energy can be followed by a sharp line in its absorption spectrum at the wavelength λ given by $\Delta E = hc/\lambda$ (see Chapter 1). In a molecule consisting of various atoms the transition from the ground to the excited state can take place by the absorption of light of varying amounts of energy quanta; the sharp line of the atomic absorption spectrum then is replaced by a broad absorption band. In individual atoms absorption and emission take place at the same wavelength, while in a whole molecule the absorption and emission spectrum do not coincide; the peak of the emission spectrum is at a longer wavelength than the peak of the corresponding absorption spectrum (Fig. 4–2a).

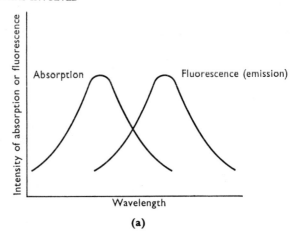

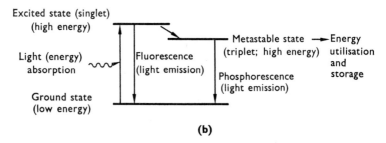

Fig. 4–2   Spectra of light absorption and emission.

4.1   Time spans involved; fluorescence and phosphorescence

A molecule, say of chlorophyll, in an electronically excited state can revert to the ground state in a number of ways. It can dissipate some of the acquired energy as heat and emit a photon back into space (Fig. 4–2 (b)). This phenomenon is called *fluorescence*. The wavelengths of a fluorescence spectrum are always longer than the wavelengths of the corresponding absorption spectrum. Chlorophyll *a* extracts, for example, absorb in blue and red regions of the spectrum but fluoresce only in the red; the maximum intensity of fluorescence occurs at 668 nm compared to the wavelength of maximum absorption at 663 nm. The average time a molecule has to spend in the excited state between absorption and emission is known as the natural lifetime of the excited state; its duration depends on the electronic properties of the excited state. The average fluorescence lifetime is of the order of 10^{-9} s.

A second route by which an excited molecule can lose its energy is by transfer from its original excited singlet state into a metastable triplet state with a much longer lifetime (of the order of milliseconds). From the meta-

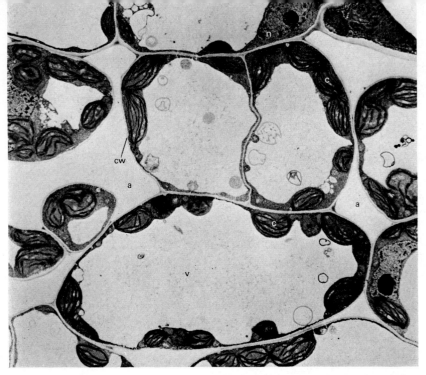

Plate 1 Thin section of spinach mesophyll cells showing chloroplasts (c) in cytoplasm extending around the inside of the cell wall (cw). n = nucleus, v = vacuole, a = air space between cells allowing easy diffusion of gases to chloroplasts. (× 2500) (Courtesy A. D. Greenwood, Department of Botany, Imperial College, London.)

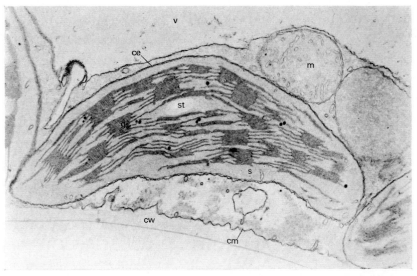

Plate 2 Section of a chloroplast in the cytoplasm of a spinach leaf cell. ce = chloroplast envelope, g = granum consisting of stacks of thylakoids, s = stroma, st = starch granule in chloroplast, cm = cytoplasmic membrane, cw = cell wall, m = mitochondrion, v = vacuole. (× 20 000) (Courtesy A. D. Greenwood.)

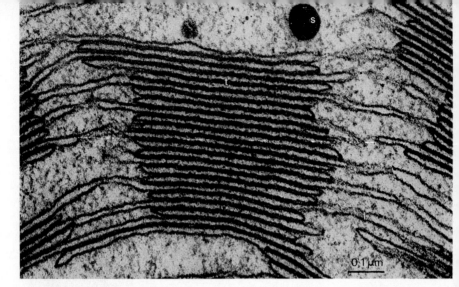

Plate 3 A single granum within a chloroplast showing stacks of thylakoids (t) and interconnecting stroma lamellae (l) between granal stacks. s = lipid droplet in the stroma. (× 103 000) (Courtesy A. D. Greenwood.)

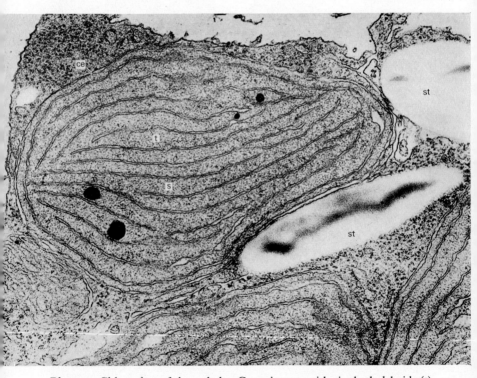

Plate 4 Chloroplast of the red alga *Ceramium* sp. with single thylakoids (t) lying nearly parallel to one another in the stroma (s). ce = chloroplast envelope, st = starch granule outside chloroplast. Dark spots in chloroplast are lipid droplets. (× 4400) (Courtesy A. D. Greenwood.)

stable triplet state the molecule can revert to the natural ground state by emitting a photon at a longer wavelength. This weak emission is known as *phosphorescence*. Phosphorescence is slow enough to be observed by the eye even when the exciting light is turned off. Due to their longer lifetime, lower energy, and magnetic moment (since the excited electron and its partner have parallel spins) the *triplet excited states* are of importance in photochemistry. The probability of an excited chlorophyll molecule reacting with another molecule is very much higher in the triplet(phosphorescent) state than in the singlet (fluorescent) state. The existence of triplet states has been shown in chlorophyll dissolved in organic solvents but not yet clearly demonstrated in vivo.

4.2   Energy transfer or sensitized fluorescence

The phenomenon of sensitized fluorescence involves the interaction of two molecules that may be separated in solution by many molecules of the solvent. In this type of energy transfer, two kinds of pigments are dissolved in the same solvent and the solution is illuminated with light of such wavelength that can be absorbed by only one of the pigments called the *donor*. The wavelength of light emitted from the solution, however, corresponds to the fluorescence spectrum of the second pigment molecule, the *acceptor*. The energy of excitation of the donor molecule is transferred by resonance to the acceptor molecule. One of the requisites for this type of energy transfer is that the fluorescent state of the donor molecule must have an energy

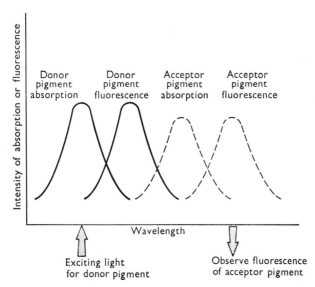

Fig. 4-3   Relationship between absorption and fluorescence spectra of donor and acceptor pigments.

greater than or equal to the fluorescent state of the acceptor molecule, that is, the fluorescent band of the donor molecule should overlap with the absorption band of the acceptor (Fig. 4–3). The quanta taken up by accessory pigments in many algae are transferred either wholly or partly to chlorophyll *a* by sensitized fluorescence. In the green algae chlorophyll *a* fluorescence is observed when light is absorbed by chlorophylls *a* and *b* and also by the carotenoids. Chlorophyll *a* has its fluorescence at the longest wavelength so that the migration of energy is always from the other excited chloroplast pigments to chlorophyll *a*.

From studies of chlorophyll *a* fluorescence and quantum yields of photosynthesis in *Chlorella* it was shown that in this alga the transfer of excitation energy from chlorophyll *b* to *a* is 100% efficient while the energy transfer from the carotenoids to chlorophyll *a* is only 40% efficient. The close assembly of various pigment molecules in the lamellae is necessary for efficient energy transfer. The three principal modes of light emission are shown in Fig. 4–2 (b).

4.3   Emerson effect and the two-light reactions

A plot showing the efficiency of photosynthesis (measured as O_2 evolution or CO_2 fixation) by monochromatic light as a function of the wavelength of light is known as the *action spectrum* of photosynthesis. For photochemical reactions involving a single pigment the action spectrum has the same general shape as the absorption spectrum of the pigment. If P molecules of O_2 per second are evolved from a system which absorbs I quanta of monochromatic radiation per second, then the ratio P/I, (Φ), is called the quantum yield or *quantum efficiency* of photosynthesis. The reciprocal of the quantum yield $[1/\Phi]$ which gives the number of quanta required to liberate one molecule of O_2 is usually called the *quantum requirement* of photosynthesis. Although values ranging from 4 to 12 have been mentioned for the quantum requirement by various workers in the field the more widely accepted value is about 8.

Emerson and associates at the University of Illinois in the 1940s studied the action spectra of photosynthesis for various algae by measuring the maximum quantum yield of photosynthesis as a function of the monochromatic light used to illuminate the algae. They found that the most effective light for photosynthesis, in *Chlorella*, was red (650 to 680 nm) and blue (400 to 460 nm), those colours that are most strongly absorbed by chlorophyll. The photosynthetic efficiency of a quantum absorbed at 680 nm was about 36% more than that of a quantum absorbed at 490 nm.

The quantum yield of photosynthesis decreased very dramatically with increasing wavelength beyond 685 nm even though chlorophylls still absorb light at these wavelengths. This fact, the so-called *red drop* in photosynthesis, could not be explained at that time. However, Emerson and co-workers later showed that the amount of photosynthesis in far red

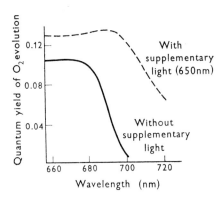

Fig. 4–4 Efficiency of photosynthesis (quantum yield) in the green alga *Chlorella* at different wavelengths of light. On adding supplementary light the quantum yield is enhanced at wavelengths above 680 nm—the Emerson enhancement effect. (Redrawn from EMERSON *et al.* (1957) *Proc. Natl. Acad. Sci. U.S.*, **43**, 133.)

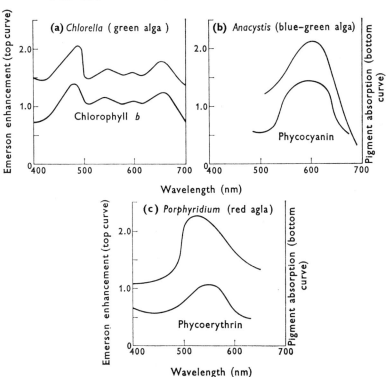

Fig. 4–5 Action spectrum of the Emerson effect in different algae (upper curve in each case) correlated with the absorption of the accessary pigments of the same algae (lower curve). (a) *Chlorella* containing chlorophyll *b*; (b) *Anacystis* containing phycocynanin; (c) *Porphyridium* containing phycoerythrin. (Redrawn from EMERSON and RABINOWITCH (1960), *Plant Physiol.*, **35**, 477.)

light (wavelengths greater than 685 nm) could be increased considerably by a supplementary beam of red light (about 650 nm) (see Fig. 4–4). In fact the total amount of photosynthesis carried out in the presence of a mixture of far red light and red light was greater than the sum of the amounts of photosynthesis carried out in separate experiments with the individual beams of light. This increase of the photosynthetic efficiency of far red light in the presence of a supplementary beam of lower wavelength is known as the *Emerson enhancement effect*. Experimental results of this effect are shown for three plants with different accessory pigments in Fig. 4–5. What is actually measured is O_2 evolution. It will be noticed that the simultaneous illumination of the plant by light of two different wavelengths results in greater O_2 evolution than the sum of that yielded by the two wavelengths applied separately. In other words, enhancement has occurred. Thus plant photosynthesis involves two light reactions; one sensitized by chlorophyll *a* (System I) and one sensitized by accessory pigments (System II).

Important subsequent work by Meyers and French in 1960 showed that Light System I and Light System II need not be applied simultaneously to get optimum photosynthesis but that they may be given alternately with a short dark period of a few seconds between them. This indicated that the two-light reactions could store their photochemical products for a short time before reacting with the electron transfer chain.

Experimental evidence and theoretical postulates in support of the two-light reaction hypothesis followed. The difference in electrode potential (ΔE) between the reactants and the final products in the photosynthetic reaction is 1.25 V (ΔE of CO_2 – glucose couple = – 0.43 V; ΔE of H_2O – O_2 couple = + 0.82 V) which is equivalent to 48.5×10^4 J (116 kcal). This

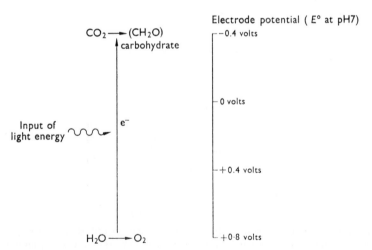

Fig. 4–6   Redox potentials of the overall reaction of photosynthesis.

energy gap is greater than that which can be overcome by the absorption of even two quanta of orange-red light (18.37×10^4 J/mole). In order to evolve one molecule of O_2, four electrons are required (pp. 13, 43). Thus if a quantum requirement of 8 is accepted as the value for photosynthesis (p. 30) two light reactions would be needed for each O_2 evolved since one quantum of light can only activate one electron (8 quanta required per O_2/4 electrons per $O_2 = 2$ quanta per electron). In 1960 Hill and Bendall

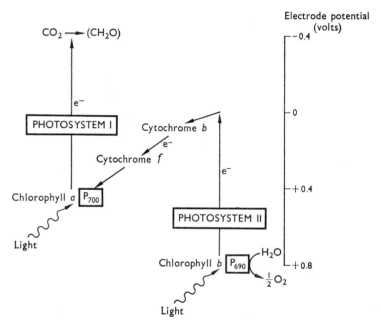

Fig. 4–7 The two-light reaction scheme of photosynthesis. (Concept of HILL and BENDALL (1960), *Nature*, **186**, 136.)

at Cambridge put forward the idea that the two-light reactions should be in series (and not in parallel) with cytochromes *b* and *f* acting as electron carriers in the 'dark' reaction which connects the two photosystems. This is shown in more detail in Fig. 4–7 and is elaborated further in Chapter 5. The most conclusive evidence for two separate photosystems came from a series of difference spectroscopy studies initiated by Duysens and by Kok and elaborated further by Witt. In this type of measurement the change in absorbancies of various constituents of a cell suspension are studied by illumination with monochromatic light of varying wavelengths. A schematic diagram of a difference spectrophotometer is shown in Fig. 4–8. By following the qualitative and quantitative changes in the difference spectrum of individual photosynthetic reaction components, e.g. cytochromes, the role of some of these components in the electron transfer pathway can be inferred. For example, Duysens illuminated a suspension of the red alga *Porphyridium* in the presence of DCMU (a synthetic weed killer which inhibits oxygen evolution) and found an accumulation of cytochrome *f* in the oxidized form. When the experiment was repeated with unpoisoned alga there was no change in the cytochrome *f* spectrum. Duysens concluded that cytochrome *f* was an intermediate in the reactions accompanying oxygen evolution (see also Chapter 5). By illuminating algae suspensions

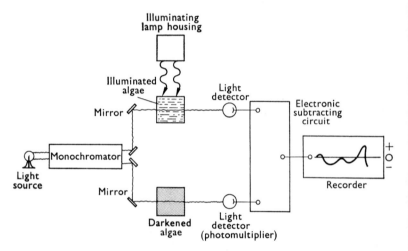

Fig. 4–8   Diagram of a difference spectrophotometer to measure the absorption of pigments.

with brief flashes of light Kok was able to identify and characterize a special type of chlorophyll *a* which he called P_{700}. P_{700} is a trace constituent of chloroplasts with an absorption maximum around 700 nm. Chemical titration showed P_{700} to be a one-electron acceptor molecule with a potential of +0.45 V. It is reversibly bleached in light; the bleaching corresponds to an oxidation of the pigment. Normal chloroplasts contain one P_{700} molecule for every 250–300 chlorophyll molecules. According to Kok, P_{700} is the 'trap' pigment in which the light energy is captured and utilized for primary photochemical reactions (see Fig. 3–8). The flash-light experiments of Witt and co-workers suggested that light displaces an electron from P_{700} to an acceptor and an electron simultaneously moves from cytochrome *f* to P_{700}. Similar studies were conducted to decide the role of chlorophyll *b*, various types of chlorophyll *a*, the accessory pigments and other constituents in the two-light reactions. An entirely new approach was devised by Levine, Bishop and others who studied the photosynthetic efficiencies of mutants of algae lacking one or more components of the photosynthetic system in order to determine the role of these constituents in the two-light reactions of photosynthesis.

The present state of our knowledge can be summarized as follows. All photosynthetic pigments in plants and algae are distributed into two systems, 'System I' (PSI) and 'System II' (PSII) which are triggered by light reactions I and II respectively. PSI is weakly fluorescent, contains a relatively lower proportion of chlorophyll b and three types of chlorophyll a, viz. a_{670}, a_{680} and a_{695}. PSI also contains the light-trapping pigment P_{700}. PSII is relatively strongly fluorescent, contains a higher proportion of chlorophyll b and two forms of chlorophyll a, viz. a_{670} and a_{680}. The light-trapping agent in this system is a form of chlorophyll absorbing at 690 nm, P_{690}. The two pigment systems sensitize two consecutive photochemical steps and probably function in series rather than in parallel. Details of the scheme of operation of the two systems are given in the next chapter.

4.4 Experimental separation of the two photosystems

In recent years various attempts have been made to physically separate PSI and PSII from chloroplasts. Chloroplasts are fragmented further by the addition of detergents like digitonin or Triton X-100, by sonic vibration or by mechanical extrusion under high pressure (about 800 atmospheres) through a French Press. The fragments are then separated by a combination of differential and density gradient centrifugation and chromatography technique. Some of the preparations thus obtained had a very high chlorophyll a to chlorophyll b ratio and a high P_{700} content (as high as 1 P_{700} per 16 chlorophylls) suggesting they are highly enriched in PSI constituents. However, a complete separation of the active pigment systems has not yet been achieved. It seems to be very difficult to isolate a PSII particle free of PSI particles. Attempts have been made to localize the positions of PSI and PSII particles in stroma and grana lamellae but this work is still in its infancy and is fraught with difficulties.

Photosynthetic phosphorylation

In Chapter 2 we elucidated the idea of photosynthesis involving both light and dark phases in the fixation of CO_2. The recognition and experimental demonstration of these two phases was an important step towards the modern understanding of the CO_2 fixation process. This was not possible until Arnon, Allen and Whatley in 1954 were able to isolate chloroplasts from spinach leaves which were capable of carrying out complete photosynthesis, i.e. fixing CO_2 to the level of carbohydrate (see Chapter 3 for the preparation technique of chloroplasts). They were physically able to separate the light and dark phases and to show the light-dependent formation of ATP and $NADPH_2$, which then acted as the energy sources for the subsequent dark fixation of CO_2. This is summarized in the familiar diagram below (Fig. 5–1).

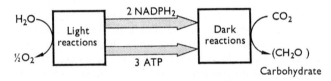

Fig. 5–1 Major products of the light and dark reactions of photosynthesis.

The light phase, which occurs subsequent to the initial light reactions discussed in the previous chapter, involves biochemical reactions with life times of 10^{-1} to 10^{-5} s. The initial light reactions have of course much shorter lifetimes—down to 10^{-9} s. The biochemical events of the light phase results in (i) the production of the strong reducing agent, $NADPH_2$, (ii) the accompanying evolution of O_2 as a by-product of the splitting of H_2O and (iii) the formation of ATP which is coupled to the flow of electrons from H_2O to NADP.

In this chapter we shall discuss how these reactions are thought to occur, what evidence there is for making such assumptions and what compounds are involved in the sequence of electron transport reactions.

5.1 Reduction and oxidation of electron carriers

The production of $NADPH_2$, ATP and O_2 in the lamellae of the chloroplast involves the transfer of electrons through a chain of electron carriers. This electron transfer requires that each of the carriers in turn becomes reduced and oxidized in order that the energy in the electron can be passed along the chain. Reduction simply means the adding of an electron while oxidation implies the removal of an electron from a compound. Whenever an electron is exchanged between two compounds one is oxidized and the other reduced. Almost every such exchange is accompanied by the release or absorption of energy. It makes no difference whether we think of the energy as arising out of the pull exerted on the electron by 'oxidizing power' or the push exerted by 'reducing power'.

Often, though not invariably, an electron travels in company with a proton, i.e. as part of a hydrogen atom. In that case oxidation means removing hydrogen and reduction means adding hydrogen. Thus NADP is reduced to $NADPH_2$, and CO_2 to carbohydrate, by the addition of hydrogen atoms.

The oxidation-reduction ('redox') potentials of biological electron carriers are expressed on a voltage scale at biological pH's which indicates that the $H_2O \rightarrow O_2$ couple is very oxidizing, with a positive potential of $+0.82$ V while the $H^+ \rightarrow H_2$ (gas) couple is very reducing with a negative potential of -0.42 V. Most biological electron transfer reactions occur between these two extremes (see Table 3). We shall see later that, in fact, the process of photosynthetic electron transport takes place at between $+0.8$ V and -0.4 V. In order to reach these extremes of redox potential light energy is required.

Table 3 The oxidation-reduction (redox) potentials of some reactions and chloroplast components. The compounds change from the oxidized to the reduced forms at the redox potentials listed at pH 7.

Chloroplast components	Redox potential (volts)
H_2O/O_2	$+0.81$
Chlorophyll a	$+0.43$
Plastocyanin	$+0.37$
Cytochrome f	$+0.37$
Plastoquinone	0.0
Cytochrome b_6	-0.07
$NADPH_2$	-0.34
Ferredoxin	-0.43
H^+/H_2	-0.42

5.2   Two types of photosynthetic phosphorylation

Photosynthetic phosphorylation is the production of ATP in the chloroplast by light-activated reactions; it can take place via two systems, non-cyclic and cyclic. In non-cyclic photophosphorylation ATP is generated in an 'open' electron transfer system together with the evolution of O_2 from H_2O and the formation of $NADPH_2$ from NADP. In cyclic photophosphorylation the electrons cycle in a 'closed' system through the

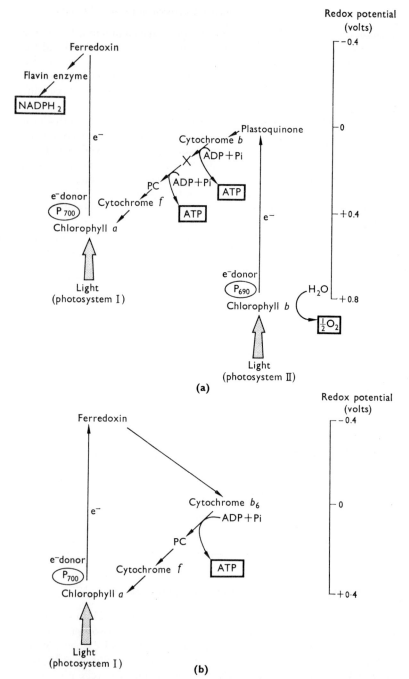

Fig. 5–2 Scheme for (**a**) non-cyclic and (**b**) cyclic photosynthetic phosphorylation.

phosphorylation sites and ATP is the only product formed. These two systems are shown in Fig. 5–2.

5.3   Non-cyclic photosynthetic phosphorylation

This is the light-requiring process in which low energy electrons are removed from H_2O, resulting in the evolution of O_2 as a by-product (Hill reaction, Chapter 2), and the transfer of these electrons via a number of carriers to produce a strong, negative reducing potential with the subsequent formation of $NADPH_2$, a reducing agent with a potential of -0.34 V. This is simply expressed as

$$NADP + H_2O \xrightarrow[\text{chloroplasts}]{\text{light}} NADPH_2 + \tfrac{1}{2}O_2$$

The carriers which have been identified include chlorophylls a and b, quinones, cytochromes b and f, plastocyanin, enzymes and ferredoxin.

ATP formation accompanies the transfer of electrons from the quinones to cytochrome f and involves side reactions which are obligatorily coupled to the electron transfer process. Thus $NADPH_2$ and ATP formation occur in the process of non-cyclic photophosphorylation where electrons are removed from H_2O and donated to NADP with the coupled formation of ATP. This overall reaction can be expressed as:

$$NADP + H_2O + ADP + Pi \xrightarrow[\text{chloroplasts}]{\text{light } (2e^-)} NADPH_2 + ATP + \tfrac{1}{2}O_2$$

This equation implies that each H_2O is split in the chloroplast membrane under the influence of light to give off $\tfrac{1}{2}O_2$ molecule (an atom of oxygen) and that the two electrons so freed are then transferred to NADP, along with the H^+'s (protons) from the H_2O, to produce the strong reducing agent, $NADPH_2$. One molecule of ATP is simultaneously formed from ADP and Pi (inorganic phosphate) so that energy is stored in the form of this high energy compound.

The $NADPH_2$ and ATP are the 'assimilatory power' required to reduce CO_2 to carbohydrate in the dark phase, which will be discussed in the next chapter. Thus 'assimilatory power' represents the initial products of the conversion of light energy into chemical energy.

A diagrammatic representation of the electron flow pattern in non-cyclic phosphorylation is given in Fig. 5–2 (a). This formulation is derived from the elegant hypothesis of Hill and Bendall in 1960. The oxidation-reduction ('redox') scale on the right shows very clearly the redox potential of all the electron carriers in the chain and implies that the sequence of electron flow depends to a large extent on their potential. This is a very neat view of electron transport and all the evidence to date indicates that it is most probably correct.

What is immediately apparent is that two different light reactions are required to raise the electrons from the level of H_2O ($+0.82$ V) to the level of NADPH$_2$ (-0.34 V)—these are designated Light Systems I and II. Each of the systems requires a different type of chlorophyll. Light System I requires chlorophyll *a* which has an absorption maximum of 663 nm (bluish-green in colour) while Light System II requires the closely related chlorophyll *b* (see Chapter 3) which has its main absorption peak at 645 nm (yellowish-green in colour). Chlorophyll *a* occurs in all plants, while chlorophyll *b*, an accessory pigment, is present only in higher plants— blue-green algae may contain phycocyanin and red algae phycoerythrin as their accessory pigments for Light System II.

The first evidence for the possible involvement of two different light re- actions in photosynthesis came from the work of Emerson and co-workers from 1943 on (see also Chapter 4). They showed that the reduction of one molecule of CO_2 to carbohydrate required light of two different wave- lengths if one of the lights was *only* absorbed by chlorophyll *a*, e.g. light of wavelength greater than 680 nm. The wavelength of the second light re- quired to give efficient photosynthesis was found to correspond to that of chlorophyll *b* in higher plants and to the other accessory pigments in lower plants.

Further evidence for the requirement of light of two different wave- lengths in non-cyclic electron flow from H_2O to NADP came from experi- ments which measured the changes in oxidation and reduction state of the cytochromes in the chloroplasts. In Fig. 5–2 it is seen that cytochrome *f* is an electron carrier intermediate between Light Systems I and II. Using very sensitive spectrophotometric techniques (see Chapter 4) it is possible to measure the redox state of the cytochrome *f*, due to its specific absorp- tion peaks at 422 and 550 nm, under illumination of light of different wave- lengths being absorbed by the two Light Systems. The experimental results are shown in Fig. 5–3. Cytochrome in the chloroplast is naturally in the reduced state in the dark and thus if isolated chloroplasts are illuminated

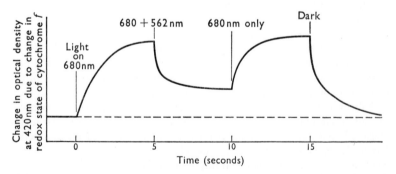

Fig. 5–3 Oxidation and reduction of cytochrome *f* in the red alga *Porphyri-dium*. Increase in OD$_{420}$ is due to oxidation and decrease in OD$_{420}$ is due to reduction of cytochrome *f*. Light of 680 nm is absorbed by Photosystem I (chlorophyll *a*) and at 562 nm is absorbed by Photosystem II (phycoerythrin). (After DUYSENS and AMESZ (1962), *Biochim. Biophys Acta*, **64**, 243.)

with light of 680 nm wavelength, which is only absorbed by Light System I, i.e. chlorophyll *a* in the red algae *Porphyridium*, electrons are removed from cytochrome *f* and donated to ferredoxin and thence to NADP—thus the cytochrome *f* becomes oxidized. Then if light at 562 nm (absorbed by Light System II, phycoerythrin in *Porphyridium*, but also to some extent by Light System I) is applied to the chloroplast the cytochrome *f* becomes reduced since it accepts electrons from Light System II. Note that complete reduction is not achieved since Light System I is still functioning to a limited extent in removing the electrons from the cytochrome *f*. In the dark the cytochrome *f* returns to its normal reduced state seen at the start of the experiment. This type of experiment was initiated by Duysens in 1961.

Similar experiments have recently been done to show the reduction and oxidation of cytochrome *b* in the chloroplast. This type of experiment is of great importance in localizing the site of electron carriers in the chain. The technical difficulties are great as one must be able to measure specific changes in one carrier at a time—a difficult but very rewarding technique.

Many different compounds have been isolated from chloroplasts (see Chapter 3) but the electron carriers shown in the scheme for non-cyclic photophosphorylation (Fig. 5–2) are those for which a role has been assigned so far. These compounds have been isolated and characterized chemically to a greater or lesser extent, e.g. quite a lot is known about ferredoxin and plastocyanin but relatively little about cytochromes *b* and *f*. Experiments have been devised which enable one to remove a specific carrier from the chloroplasts, e.g. by washing with water or low concentrations of detergents, organic solvent extraction or by mild sonication; a certain electron transfer reaction is thus diminished and then the re-addition of the extracted compound in its purified form restores the electron-carrying ability of the chloroplast membranes. This type of experiment has been successfully accomplished with plastoquinone, plastocyanin, ferredoxin and the enzyme (a flavoprotein) acting between ferredoxin and NADP. These types of experiments add further evidence to that accumulated in other investigations on the relative position of the electron carriers.

Elegant experiments by Levine using genetic mutants of the green alga *Chlamydomonas* have also helped consolidate the electron flow sequence of Fig. 5–2. Specific mutants of the alga have been obtained which are deficient in certain parts of the electron transfer chain, e.g. blocking electron flow between plastoquinone and cytochrome *b* or between plastocyanin and cytochrome *f*. With a knowledge of the exact site of the blockage experiments can be designed in which electrons are added and subtracted at various parts of the chain. This can be quite easily accomplished by using different types of dyes and is also used successfully in the spectrophotometric experiments mentioned earlier. These genetic and biochemical experiments which are similar in principle to those used with the mould *Neurospora* and the bacterium *Escherichia coli*, have provided confirming evidence for the non-cyclic electron flow sequences and should prove valuable in our further elucidation of photophosphorylation and CO_2 fixation.

The use of chemical inhibitors of specific biochemical reactions is a classical and fruitful approach to understanding biochemical mechanisms. Non-cyclic photophosphorylation is no exception to the advantageous use of specific inhibitors. A number of different inhibitors have been found which block specific parts of the chain, e.g. *DCMU*, a herbicide, which blocks oxygen evolution; *antimycin A*, an antibiotic, which prevents the reduction of cytochrome *f*; and *DSPD* which blocks ferredoxin-catalyzed electron flow.

We have so far said very little about the oxygen evolution end of the

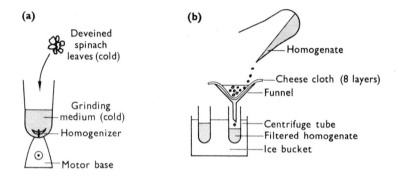

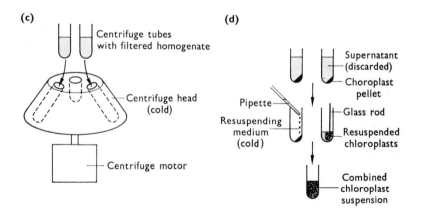

Fig. 5–4   Experiment to demonstrate photosynthetic O_2 evolution and NADPH$_2$ and ATP formation by isolated spinach chloroplasts.

chain where Light System II operates in conjunction with chlorophyll b. This is simply because we know very little about the mechanism of the splitting (or photolysis) of H_2O to give off oxygen except to say that Mn and Cl^- ions seems to have catalytic roles in the reaction(s) involved. The reaction is postulated to occur as follows:

$$H_2O \longrightarrow H^+ + OH^-$$

$$OH^- \longrightarrow \tfrac{1}{2}O_2 + H^+ + 2e^-$$

$$H_2O \longrightarrow 2H^+ + \tfrac{1}{2}O_2 + 2e^-$$

(e)

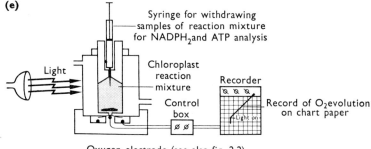

Oxygen electrode (see also fig. 2·2)

(f)

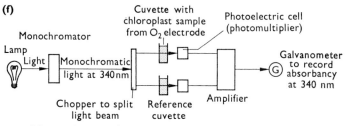

Scheme of spectrophotometer to measure $NADPH_2$ at 340 nm

(g)

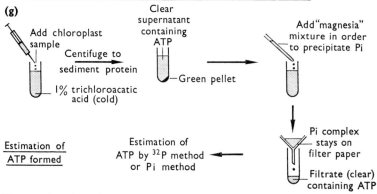

Fig. 5–4 (continued)

This great gap in our knowledge is unfortunate because of the fundamental importance of the H_2O splitting reaction. Chemically it is very important to solve but biologically even more so since all the oxygen we breathe is derived from this reaction in the chloroplast!

Lastly we must discuss the ATP formation which is coupled to the non-cyclic electron flow. In 1958 Arnon, Whatley and Allen demonstrated the obligatory coupling of ATP formation to the reduction of NADP and showed that the rate of electron flow to NADP was dependent on the presence of ADP and Pi (which is required to form ATP). How many ATP molecules are formed per $NADPH_2$ produced is an important question because of the amount of ATP required to fix CO_2 in the dark phase—2 $NADPH_2$'s and 3 ATP's are required per CO_2 molecule reduced to the level of carbohydrate. The present evidence suggests 2 ATP per $NADPH_2$. The mechanism of coupling between the electron transfer chain and the formation of ATP itself is thought to involve high-energy states or intermediates and so-called 'coupling factors' which are proteins associated with the chloroplast membrane.

Recently Mitchell in England has proposed a 'chemiosmotic' theory to account for ATP formation in membranes. This hypothesis envisages ATP synthesis occurring as the result of the recombination of positive and negative charges across the grana membrane—the initial separation of the charges being brought about by the transfer of electrons along the electron carrier chain. This idea requires that the membranes have a definite orientation and ability to bring about this charge separation; it is attractive since the chloroplast membranes are similar to other membranes involved in ATP synthesis and breakdown, e.g. mitochondria, and the theory is proposed to work for all types of membranes (see Tribe and Whittaker, 1972).

During the isolation and preparation of chloroplasts from the whole leaf the ATP formation factors are easily destroyed so that greater care must be taken in performing phosphorylation experiments than those in which only electron transport is measured. In Fig. 5–4 a sequence of diagrams shows how one isolates chloroplasts and measures the O_2 evolution, $NADPH_2$ formation and ATP formation associated with non-cyclic photophosphorylation (see also Chapter 3).

5.4   Cyclic photosynthetic phosphorylation

In this process, which requires light and chloroplasts, the only net product is ATP. The reaction was discovered in 1954 by Arnon, Allen and Whatley using isolated spinach chloroplasts and by Frenkel using chromatophores isolated from photosynthetic bacteria. It may be very simply represented by the following equation:

$$\text{ADP} + \text{Pi} \xrightarrow[\text{chloroplasts}]{\text{light}} \text{ATP}$$

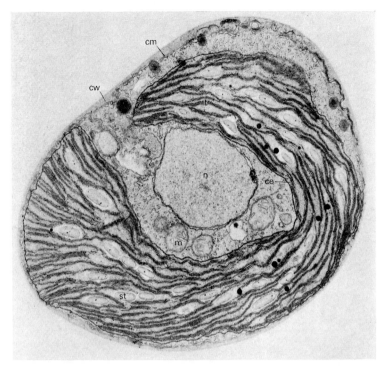

Plate 5 The green alga *Coccomyxa* sp., which is a symbiont within a lichen, showing a single cup-shaped chloroplast within the cell. The thylakoids (t) are in groups of three. ce = chloroplast envelope, st = starch granule within chloroplast, cw = cell wall, cm = cytoplasmic membrane, n = nucleus, m = mitochondrion. (× 12 400) (Courtesy H. Bronwen Griffiths, Department of Botany, Imperial College, London.)

Plate 6 (left) Bean chloroplasts isolated in buffered sucrose media showing chloroplasts with intact thylakoids but without a chloroplast envelope—'Class II' as defined in chapter 3. (×8300) (Courtesy A. D. Greenwood and Rachel Leech.)

Plate 7 (above) Surface view of chromium-shadowed spinach chloroplast thylakoid (t). Regular repeating array of particles (p) called quantasomes, within the lamella is readily seen. The exact nature and function of these particles is uncertain. (Courtesy R. B. Park, Department of Botany, University of California, Berkeley.)

Only a cyclic electron flow involving Light System I is required in order to produce ATP. Figure 5–2 (b) suggests how this may occur. Under the influence of an input of light an electron is removed from chlorophyll a in its excited state and donated to ferredoxin which becomes reduced. The reduced ferredoxin then, instead of transferring its electron to NADP as in the case of non-cyclic electron flow, donates its electron to cytochrome b and thence through the electron transport chain back to chlorophyll a. Thus the electron undergoes a cyclic flow and the only measurable product is ATP which is formed by a coupling mechanism, probably similar to that involved in non-cyclic photophosphorylation even though the electron carriers may not be identical. The number of molecules of ATP formed per electron transferred is so far undetermined because of the difficulty in measuring the number of electrons cycling around the chain in a given time—the number of ATP's formed in a given time is, on the other hand, relatively simple to measure.

We can thus see that ferredoxin may play a central role in photosynthesis. It can donate electrons in a non-cyclic system to NADP in order to produce the strong reducing power in the form of $NADPH_2$ needed for CO_2 reduction; or it can donate electrons back into the electron transfer chain in a cyclic system resulting only in the formation of ATP. This ATP can be used for CO_2 fixation or for other reactions which only require ATP as their energy source, e.g. protein synthesis and the conversion of glucose to starch, both of which occur in the chloroplast. The physiological controlling mechanism associated with ferredoxin and its very low reducing potential of -0.43 V (equivalent that of H_2 gas) have stimulated much research into the physicochemical and biochemical properties of this unique protein. A model for the active centre of ferredoxin is given in Fig. 5–5.

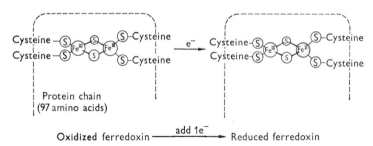

Fig. 5–5 A model for the active centre of plant ferredoxins showing the oxidized and reduced forms of the protein. (After RAO, CAMMACK, HALL and JOHNSON (1971), *Biochem. J.*, **122**, 257.)

In the chloroplast ferredoxin is probably the physiological carrier (or co-factor) involved in cyclic photophosphorylation. However, experimentally we can replace ferredoxin by vitamin K_3, FMN or any of a number of dyes. The cyclic electron flow sequence may not be exactly the same as that for ferredoxin but again the only net product is ATP.

5.5 Structure–function relationships

The light phase of the overall CO_2 fixation to the level of carbohydrate has been shown to occur in the grana lamellae (or thylakoids) of the chloroplast while the dark phase occurs in the stroma of the chloroplast. Arnon and co-workers demonstrated this in 1958 by physically separating the light and dark phase (see Table 4). Chloroplasts were illuminated in the absence of CO_2 and allowed to form large amounts of $NADPH_2$ and ATP with the concomitant O_2 evolution from the non-cyclic electron flow. The chloroplasts were then broken and the stroma separated from the grana lamellae which were discarded. In the dark, radioactive CO_2 was supplied and the enzymes in the stroma then proceeded to assimilate the CO_2 to produce the same carbohydrates that whole chloroplasts and intact green leaves manufacture.

Table 4   Carbon dioxide fixation in the dark and light by chloroplast systems, i.e. stroma (yellowish matrix) and grana (chlorophyll-containing, green membranes) (TREBST, TSUJIMOTO and ARNON (1958), *Nature*, **182**, 351).

	$^{14}CO_2$ *fixed* (*counts per minute*)
1. Stroma (dark)	4 000
2. Stroma (dark) + grana (light)	96 000
3. Stroma (dark) + ATP	43 000
4. Stroma (dark) + $NADPH_2$ + ATP	97 000

Note the equivalence of grana (light) and $NADPH_2$ + ATP, i.e. assimilatory power.

These experiments very neatly showed that all the electron carriers and enzymes required for the light-induced $NADPH_2$ and ATP formation via cyclic and non-cyclic electron flow are associated with the chloroplast membranes (grana lamellae or thylakoids). The enzymes for CO_2 fixation itself occur in the yellowish-coloured, amorphous stroma of the chloroplast. The task for the electron microscopist is now to localize the electron carriers on the lamellae themselves more exactly.

Carbon dioxide fixation 6

In the previous chapter we have seen that $NADPH_2$ and ATP are produced in the light phase of photosynthesis. The fixation of CO_2 then takes place in the dark phase using the 'assimilatory power' of $NADPH_2$ and ATP. In this chapter we shall examine in some detail the reactions involved in the reduction of CO_2 to the level of carbohydrate since the reaction mechanisms and experimental techniques, so clearly worked out by Calvin and his co-workers from 1946 on, are some of the most important in modern biology. For his work in elucidating the path of carbon in photosynthesis Calvin received the Nobel Prize for Chemistry in 1961.

6.1 Experimental techniques

When the long-lived isotope of carbon, ^{14}C, became available in 1945 its use, coupled with two-dimensional paper chromatography developed a few years earlier, enabled experiments to be devised to investigate the pathway of photosynthetic $^{14}CO_2$ fixation. The unicellular green algae *Chlorella* and *Scenedesmus* were used in the experiments because of their biochemical similarity to higher green plants and because they could be grown under uniform conditions and subsequently very quickly killed in the short-time experiments used.

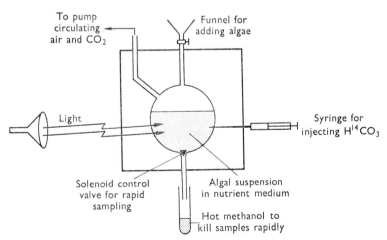

Fig. 6-1 Schematic representation of apparatus for studying $^{14}CO_2$ fixation in photosynthesizing algae. (Diagrammatic; from CALVIN and BASSHAM, 1962.)

Three main types of experiments were performed to obtain the evidence required to postulate the detailed reactions of the cycle:

(a) Exposure of the photosynthesizing algae to $^{14}CO_2$ for different lengths of time. At the shortest times only the initial products will be radioactive. In this way phosphoglyceric acid (PGA) was identified as the primary carboxylation product; end-products such as sucrose became radioactive much more slowly.

(b) Determination of the position of radioactivity within the labelled compounds. In this way the details of the interconversions of sugar phosphate to regenerate the specific sugar phosphate which accepts the $^{14}CO_2$ molecule and the mechanism of synthesis of sugars and other compounds were worked out.

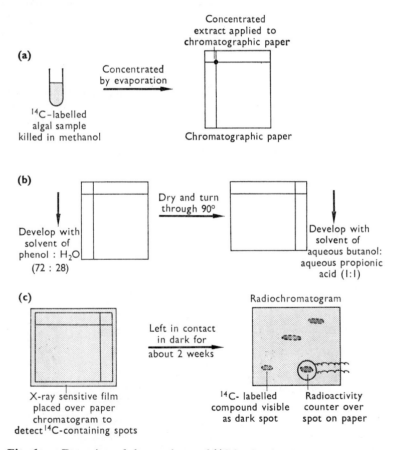

Fig. 6–2 Detection of the products of $^{14}CO_2$ fixation by algae after brief periods of illumination by the use of paper chromatography and autoradiography.

(c) Alteration of the external conditions, e.g. changing from light to dark, or changing from high to very low CO_2 concentrations, to see whether the cycle intermediates behave in a predictable manner.

The techniques employed are pictured in Figs. 6–1, 6–2 and 6–3. Figure 6–1 is a diagram of the apparatus used for obtaining extracts of algae which have been photosynthesizing $^{14}CO_2$. The algae are suspended in a nutrient medium through which air and CO_2 are bubbled with the pH of the whole suspension is maintained constant. The control valve allows rapid removal of samples into methanol which immediately stops all reactions from proceeding further. Figure 6–2 illustrates the further handling of the inactivated algae. The killed algal extract is concentrated by vacuum and then applied directly to a chromatogram paper which is developed with different solvents in two directions at right angles to each other. The radioactivity of the compounds which are separated by this two-dimensional chromatography is measured—their location being known from radioautograms of the kind shown in Fig. 6–3. The two radioautograms in Fig. 6–3

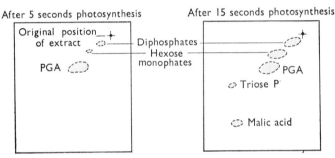

Fig. 6–3   Radioautograms of the photosynthetic products from $^{14}CO_2$ added to illuminate algae for short periods of time.

show the compounds which contain ^{14}C in extracts of *Chlorella* which had been photosynthesizing for 5 and 15 seconds. It is seen that PGA, triosephosphate and sugar phosphates are formed very rapidly; sucrose, organic acids and amino acids are only formed after longer photosynthesizing times. The combination of radioactive CO_2 and two-dimensional chromatography is seen to be a very sensitive technique for detecting and quantitatively estimating the products of photosynthesis.

6.2   The photosynthetic carbon (or Calvin) cycle

The fixation of CO_2 to the level of sugar (or other compounds) can be considered to occur in four distinct phases, as is shown in Figs. 6–4 and 6–5.

I. *Carboxylation phase* This phase is thought to consist of a reaction whereby CO_2 is added to the 5-carbon sugar, ribulose diphosphate, to form two molecules of PGA as follows:

$$CH_2OP$$
$$|$$
$$C=O$$
$$|$$
$$*CO_2 + CHOH + H_2O \xrightarrow{\text{Enzyme}}$$
$$|$$
$$CHOH$$
$$|$$
$$CH_2OP$$

$$CH_2OP \quad CH_2OP$$
$$| \qquad |$$
$$CHOH + CHOH$$
$$| \qquad |$$
$$*COOH \quad COOH$$

Ribulose diphosphate (RuDP)   2 × Phosphoglyceric acid (PGA)

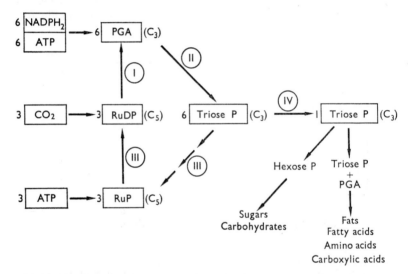

Fig. 6–4 Diagram of CO_2 fixation cycle. RuDP = ribulose diphosphate. PGA = phosphoglyceric acid. Triose P = phosphoglyceraldehyde. RuP = ribulose-5-phosphate.

The evidence in support of this scheme is shown very clearly in Fig. 6–6. On illumination RuDP and PGA increase up to a certain level which is the so-called 'steady state' level in the photosynthesizing algae. When the light is switched off the RuDP content drops immediately (as light is needed for its synthesis) while the level of PGA rises—two molecules of PGA are formed from every molecule of RuDP which disappears. Also in Fig. 6–6 is shown the effect of changing from high to very low CO_2 concentrations. A 'steady state' level is achieved at 1% CO_2 but when the CO_2

level is suddenly decreased to 0.003% (with the light still on) the level of PGA drops quickly as there is insufficient CO_2 to fix; however, the level of

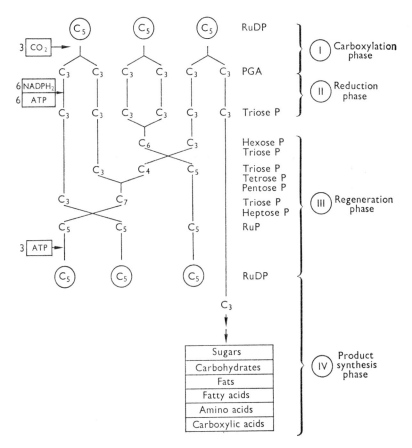

Fig. 6–5 Summary of reactions of photosynthetic CO_2 fixation. (After HALL and WHATLEY, 1967.)

RuDP increases since very little of it can be used to fix CO_2 into PGA, while it can still be formed in the light.

II. *Reduction phase* PGA formed by the addition of CO_2 to RuDP is essentially an organic acid and is not at the energetic level of a sugar. In order for PGA to be convered to a 3-carbon sugar (triose P) the energy in the 'assimilatory power' of $NADPH_2$ and ATP must be used.

The reaction is in two steps, first phosphorylation, adding on a P from ATP and then reducing with $NADPH_2$, and may be summarized as follows:

$$\begin{array}{c}
\text{CH}_2\text{OP} \\
| \\
\text{CHOH}+\text{ATP}+\text{NADPH}_2 \xrightarrow[\text{Enzymes}]{} \\
| \\
\text{COOH} \\
\text{Phosphoglyceric acid} \\
\text{(PGA)}
\end{array}
\qquad
\begin{array}{c}
\text{CH}_2\text{OP} \\
| \\
\text{CHOH}+\text{ADP}+\text{Pi}+\text{NADP}+\text{H}_2\text{O} \\
| \\
\text{CHO} \\
\text{Phosphoglyceraldehyde} \\
\text{(Triose P)}
\end{array}$$

It is seen that the reducing power of $NADPH_2$ is used to change the acid group of PGA to an aldehyde group of the triose P; ATP is required to provide the extra energy in order to accomplish this step but the Pi of ATP

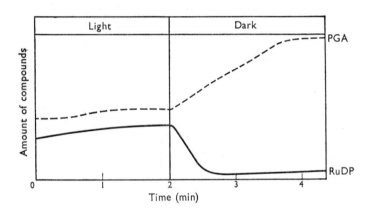

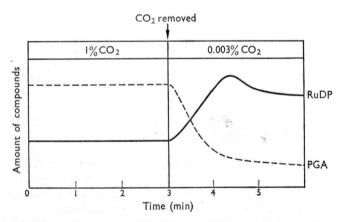

Fig. 6–6 Interconversions of RuDP and PGA, during experiments on photosynthesis. (Diagrammatic; from BASSHAM and CALVIN, in *Ruhland*, 1960.)

is not incorporated into the triose P. Both of the enzymes involved in the two steps have been shown to be present in isolated chloroplasts.

Once the CO_2 has been reduced to the level of the 3-carbon sugar, triose P, the energy-conserving part of photosynthesis has been accomplished. What is required thereafter is to regenerate the initial CO_2 acceptor molecule, i.e. ribulose diphosphate, in order for the CO_2 fixation to continue again and again (regeneration phase) and to change the triose P to more complex sugars, carbohydrates, fats and amino acids (product synthesis phase).

III. *Regeneration phase* The RuDP is regenerated for further CO_2 fixation reactions by a complex series of reactions involving 3-, 4-, 5-, 6- and 7-carbon sugar phosphates which is depicted in summary form in Fig. 6–5. The details of the reactions are not important here but can be found in the article by BASSHAM (1962). Suffice it to say that all the reactions and the enzymes involved have been studied in some detail by various groups of research workers.

IV. *Product synthesis phase* End-products of photosynthesis are considered primarily to be sugars and carbohydrates but fats, fatty acids, amino acids and organic acids have also been shown to be synthesized in photosynthetic CO_2 fixation. Many details of these synthesis reactions are known but again they do not concern us directly. What is, however,

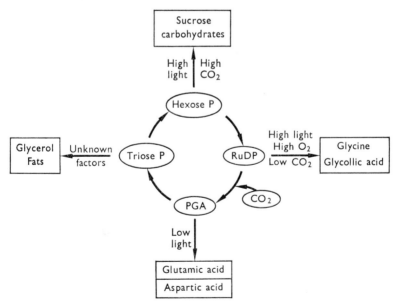

Fig. 6–7 Conditions favouring formation of secondary products in photosynthesis. (Redrawn from HILLER and WHATLEY (1967), *Advancement of Science*, p. 643.)

interesting is that the different end-products seem to be formed under different conditions of light intensity, CO_2 and O_2 concentration, as is depicted in Fig. 6–7. Much research is now being aimed at working out the synthetic reactions involved in the formation of these end products because an understanding of the reactions and the conditions favouring them may eventually enable us to induce plants to synthesize more or less sugars, fats or amino acids, by providing the required growth conditions. The practical implications are obvious if we can grow plants with controllable amounts of proteins, fats or sugars.

6.3   Structure–function relationships

In order to study CO_2 fixation in isolated chloroplasts they must be isolated rather carefully in order to preserve all the components of the reactions involved. Arnon's laboratory showed in 1954 that this could be accomplished with all the CO_2 fixation products being identified. However, the rates of fixation were only about one tenth of those observed in leaves and it was not until quite recently that Walker's and Bassham's laboratories were able, by using very careful isolation media and procedures, to obtain chloroplasts capable of CO_2 fixation rates approaching those in whole leaves. Again, the products of photosynthesis were the same as those observed by Calvin's group in whole algae and by Arnon's laboratory in the early isolated chloroplasts.

These studies emphasize that structural integrity is important in understanding how subcellular organelles such as chloroplasts do actually function. The way is now clear to study many more product synthesis reactions in isolated chloroplasts. However, basically we can still say that the light phase occurs in the grana lamellae or membranes and that the dark phase occurs in the stroma or soluble part of the chloroplast.

6.4   Energetics of CO_2 fixation

If we look at the overall and constituent equations of photosynthesis again, we will be able to examine the energy conserving and energy expending parts of the carbon fixation cycle.

The general equation for formation of glucose can be represented by:

(i) $CO_2 + H_2O \rightarrow [CH_2O] + O_2$ $\Delta G = +48 \times 10^4$ joules (114 kcal)

This means that 48×10^4 joules of energy are required to fix one mole of CO_2 to the level of glucose. This large positive value thus requires that a large amount of energy must be added.

We have already seen that this energy is derived from the light phase of photosynthesis and can be represented by the 'assimilatory power' of $NADPH_2$ and ATP. In order to fix one CO_2 molecule, two molecules of $NADPH_2$ and three of ATP are required (see Figs. 6–4 and 6–5).

The energy present in the $NADPH_2$ and ATP can be represented as follows:

(ii) $2NADPH_2 + O_2 \rightarrow 2NADP + 2H_2O$

$$\Delta G = -44 \times 10^4 \text{ joules } (-105 \text{ kcal})$$

(iii) $3ATP + H_2O \rightarrow 3ADP + 3Pi$

$$\Delta G = -9.2 \times 10^4 \text{ joules } (-22 \text{ kcal})$$

This energy is sufficient to reduce one CO_2 molecule to the level of glucose with about 5×10^4 joules (13 kcal) to spare (ii + iii − i).

The constituent parts consist of

	ΔG (joules)	ΔG (kcal)
(i) $CO_2 + H_2O$ $\rightarrow [CH_2O] + O_2$	$+48 \times 10^4$	$+114$
(ii) $2NADPH_2 + O_2$ $\rightarrow 2NADP + 2H_2O$	-44×10^4	-105 (2×52.5)
(iii) $3ATP$ $\rightarrow 3ADP + 3Pi$	-9.2×10^4	-22 (3×7.3)
	-5×10^4 joules (excess)	-13 kcal (excess)

In summary we can present the equation as:

$$CO_2 + H_2O + 2NADPH_2 + 3ATP \longrightarrow$$
$$[CH_2O] + O_2 + 2NADP + 3ADP + 3Pi$$

Thus we see that photosynthesis is essentially a reductive process since most ($44 \times 10^4 / 53.2 \times 10^4 = 83\%$) of the energy required to fix a molecule of CO_2 is derived from the strong reducing agent, $NADPH_2$, which has a redox potential of -0.34 V. The redox potentials of sugars can be thought to be approximately -0.43 V so the ATP is required to fix the CO_2 to this lower redox value.

We know that the redox potential of $H_2O \rightarrow O_2$ is $+0.82$ V so the overall change in redox potential is 1.25 V ($+0.82$ to -0.43 V). This can be converted into terms of energy (calories) using the equation

$$\Delta G = -nF \, \Delta E$$
$$= -(4)(9.7 \times 10^4)(1.25) = 48.5 \times 10^4 \text{ joules (116 kcal)}$$

where n = number of electrons ($=4$ electrons per molecule of oxygen)
        F = the Faraday ($=9.7 \times 10^4$ J per volt equivalent)
        ΔE = difference in redox potential

It is seen that this ΔG value is very close to that for fixing one molecule of CO_2 (48×10^4 joules (114 kcal); equation above) and shows quite nicely the interconversion of energy in terms of calories and redox potentials.

Lastly, we can discuss the quantum efficiency of CO_2 fixation. Each quantum of red light at 662 nm contains 17.85×10^4 joules (42.6 kcal) of energy. Thus at least three $(48 \times 10^4/17.85 \times 10^4 = 2.7)$ quanta of 662 nm light will be required for one CO_2 molecule to be fixed. However, experimentally it is found that 8–10 quanta of absorbed light are required for each molecule of CO_2 fixed or O_2 evolved. From our knowledge of non-cyclic photosynthetic phosphorylation we deduce that there are two different light reactions required to reduce NADP with the electrons from H_2O

$$2NADP + 2H_2O \xrightarrow[\substack{\text{2 light reactions} \\ \text{chloroplasts}}]{4e^-} 2NADPH_2 + O_2$$

Thus we need at least 8 quanta (4 quanta per 4e's (one O_2 molecule) $\times$ 2 light reactions) to reduce $NADPH_2$ and produce the necessary ATP at the same time.

Nevertheless photosynthetic CO_2 fixation itself is only about 30% efficient (2.7 quanta/8–10 quanta) as we can measure it. Taken in conjunction with an overall efficiency of 1–2% for whole plants capturing and utilizing sunlight (see Chapter 1) this reinforces the concept that energy exchanges are invariably very wasteful processes.

Bacterial Photosynthesis

7.1 Classification of photosynthetic bacteria

Photosynthetic bacteria are typically aquatic micro-organisms inhabiting marine and freshwater environments like moist and muddy soil, stagnant ponds and lakes, sulphur springs, etc. There are three major types.

1. *Green sulphur bacteria* (Chlorobacteriaceae) which grow by utilizing hydrogen sulphide, or in some cases thiosulphate, as electron donor, e.g. *Chlorobium.*
2. *Purple sulphur bacteria* (Thiorhodaceae) which can use hydrogen sulphide as photosynthetic electron donor, e.g. *Chromatium.*
3. *Purple non-sulphur bacteria* (Athiorhodaceae) which are unable to use hydrogen sulphide and depend on the availability of simple organic compounds like alcohols and acids as electron donors, e.g. *Rhodomicrobium, Rhodopseudomonas* and *Rhodospirillum.*

When grown photosynthetically all three types are strict anaerobes, i.e. grow in the complete absence of oxygen. They cannot use water as a substrate and they do not evolve oxygen during photosynthesis.

Photosynthetic pigments and apparatus The pigment systems of photosynthetic bacteria are slightly different from those of plants and algae. The chlorophyllous pigments of bacteria are called *bacteriochlorophylls*; four classes of bacteriochlorophylls have been characterized. Purple bacteria contain a single kind of chlorophyll, either bacteriochlorophyll *a* or *d*. Green bacteria contain bacteriochlorophylls *a* and either *b* or *c*. The principal carotenoids of photosynthetic bacteria are also slightly different chemically from the algal carotenoids. The nature of some of the photosynthetic pigments found in bacteria are given in Table 5. The absorption spectra of some of the bacteria are shown in Fig. 7–1.

The photosynthetic apparatus in purple and green bacteria are of morphologically different types and both types are distinct from the photosynthetic unit found in chloroplasts. The action spectra of bacteriochlorophyll fluorescence in purple bacteria indicate that light energy absorbed by carotenoids and shortwave bacteriochlorophyll bands is transferred to the longest wavelength bacteriochlorophyll (which absorb at 870 and 890 nm) before being used for photosynthesis. From measurements of substrate (carbon) assimilation and photosynthetic phosphorylation by suspensions of purple bacteria, during flashing light experiments, Clayton has estimated that the bacterial photosynthetic unit contains 30 to 50 bacteriochlorophyll molecules. When cells of purple bacteria are disrupted they release a class of subcellular particles containing all the photosynthetic pigments. These pigment-bearing particles can be isolated by the technique of differential

Table 5 Relationship between chlorophylls in bacteria and chlorophyll a in plants (structure on p. 21).

Type of chlorophyll	R_1	R_2	R_3	R_4	Absorption maxima, nm		Occurrence
					In organic solvents	*In cells*	
Chlorophyll a	—CH=CH$_2$	O=C, O—CH$_3$	Phytyl ester (C$_{20}$H$_{39}$O—)	—H	420, 660	435, 670–695 (several forms)	All oxygen-evolving photosynthetic systems
Bacterio-chlorophyll a	O=C, CH$_3$	O=C, O—CH$_3$	Phytyl ester	—H	365, 605, 770	Red bands at 800, 850 and 890	Purple and green bacteria
Bacterio-chlorophyll c (chlorobium chlorophyll)	OH, C—CH$_3$, H	—H	Farnesyl ester (C$_{15}$H$_{25}$O—)	—CH$_3$	432, 660	Red band at 760	Green bacteria
Bacterio-chlorophyll b	Structure unknown				368, 583, 795	Red band at 1017	Found in a strain of *Rhodopseudomonas*

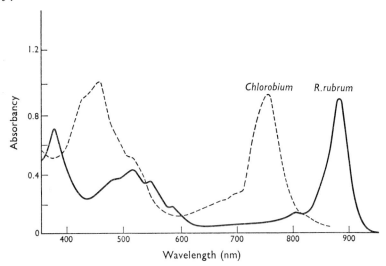

Fig. 7-1   Absorption spectra of green (*Chlorobium*) and red (*Rhodospirillum rubrum*) photosynthetic bacteria.

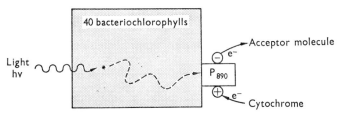

Fig. 7-2   Diagrammatic representation of the photosynthetic unit in bacteria. One photon of light reacts in a unit of 40 bacteriochlorophyll molecules containing one P_{890} reaction centre.

centrifugation. When examined by electron microscopy after staining the particles appear like spherical bodies 30 to 100 nm in diameter and are called *chromatophores*. Each chromatophore contains several photosynthetic units. They are probably derived from the external cytoplasmic membrane by extensive invaginations (infolding) of the membrane.

Duysens showed in 1952 that the absorption spectrum of bacteriochlorophyll in purple bacteria is changed reversibly by illumination. The change corresponds mainly to a bleaching (oxidation) of the long wave absorption band of bacteriochlorophyll; at 890 nm in *R. rubrum* and *Chromatium*, and at 870 nm in *R. spheroides*.

By treating chromatophores of *R. spheroides* with detergents and then illuminating them in oxygen it is possible to destroy the light-harvesting bacteriochlorophyll molecules while still keeping the light-reacting component intact. The 870 nm absorption band of such specially prepared chromatophores is oxidized reversibly by light. This special light-reacting component in *R. spheroides* is called P_{870} and its role is similar to that of P_{700} in chloroplasts. The equivalent component in *R. rubrum* and *Chromatium* is designated P_{890}. There is one P_{870} or P_{890} molecule for about 40 bacteriochlorophyll molecules which constitute the photosynthetic unit (Fig. 7–2). Difference spectroscopic studies (see Chapter 4) showed the reversible oxidation-reduction of a quinone (ubiquinone) and of a cytochrome when the light-induced changes in P_{870} were observed. The amount of light-reacting cytochrome is about 1 molecule for every 40 bacteriochlorophylls. All these support the existence of a bacterial reaction centre consisting of about 40 bacteriochlorophylls, one P_{870} or P_{890}, one cytochrome and a ubiquinone. The oxidation of cytochrome coupled to the reduction of P_{870} may be the primary photochemical act in bacterial photosynthesis.

7.2   Carbon dioxide fixation

Photosynthetic bacteria do not show an Emerson enhancement effect, so it is generally considered that there is only one major photo reaction in photosynthetic bacteria. In cell-free preparations of photosynthetic purple bacteria the dominant photosynthetic reaction is cyclic photophosphorylation (production of ATP). Particles have been prepared from green bacteria which are able to catalyze photoreduction of ferredoxin (see Chapter 5) which can subsequently be used to reduce NAD to $NADH_2$ via a flavoprotein enzyme. Thus the photosynthetic bacteria can generate both ATP (energy) and $NADH_2$ (reducing power) for the fixation of CO_2 (Fig. 7–3). The Calvin–Benson cycle (Chapter 6) enzymes are shown to be present in nearly all strains of photosynthetic bacteria studied which thus have the ability to fix CO_2 via this cycle. Recently Evans, Buchanan and Arnon have demonstrated the existence of a new route for photosynthetic CO_2 fixation in bacteria which is mediated by reduced ferredoxin leading to the synthesis of α-keto acids, e.g. pyruvate and α-ketoglutarate. Their scheme of reductive carboxylic cycle in photosynthetic bacteria is shown in Fig. 7–4. In this pathway of photosynthetic CO_2 fixation the ferredoxin (Fd)-dependent enzymes pyruvate synthase and α-ketoglutarate synthase catalyze the carboxylation of acetyl and succinyl-coenzyme A, respectively, as shown:

1. Acetyl CoA $+ CO_2 + Fd_{red}$ $\xrightarrow{\text{Enzyme}}$ Pyruvate $+ CoA + Fd_{ox}$

2. Succinyl CoA $+ CO_2 + Fd_{red}$ $\xrightarrow{\text{Enzyme}}$ α-ketoglutaric acid $+ CoA + Fd_{ox}$

The overall cycle involves the fixation of four molecules of CO_2. It is probable that in these bacteria both the reductive pentose phosphate (Calvin–Benson) pathway and the reductive carboxylic acid cycle are operating in photosynthetic CO_2 fixation.

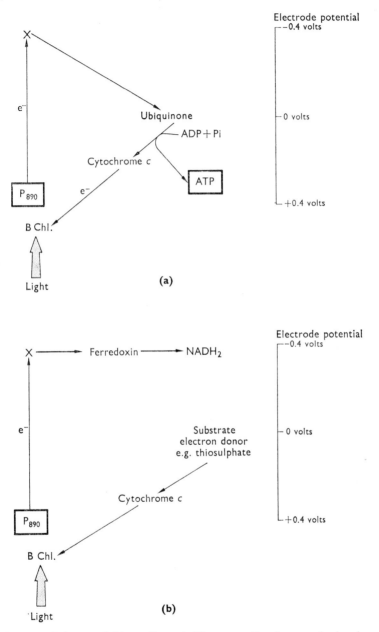

Fig. 7-3 Scheme of **(a)** cyclic and **(b)** non-cyclic photosynthetic phosphorylation in bacteria.

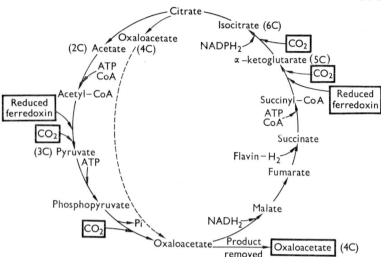

Fig. 7–4 The reductive carboxylic acid cycle of photosynthetic bacteria. Four CO_2 molecules are fixed to form one oxaloacetate molecule. (Scheme devised by EVANS, BUCHANAN and ARNON (1966), *Proc. Natl. Acad. Sci. U.S.*, **55**, 928.)

7.3 Ecological and evolutionary significance of phototrophic bacteria

Under anaerobic (O_2–free) conditions organic matter is fermented by various micro-organisms (the chemosynthetic anaerobes) which gain their energy by a substrate-linked phosphorylation. Various metabolic end products like CO_2, H_2, ethanol and simple fatty acids are formed during this process. Such compounds would accumulate if they were not removed as nutrients by other types of microbes which are unable to use oxygen as the ultimate electron acceptor in their respiratory processes. The sulphate- and nitrate-reducing bacteria are able to consume part of the end products of fermentation of the chemosynthetic anaerobes. The phototrophic bacteria (green and purple photosynthetic bacteria) derive their energy from light and are able to metabolize most of the end products of anaerobic fermentation like alcohols, acids, hydrogen and nitrogen, as well as the end products of sulphate and nitrate respirations such as H_2S and N_2. Thus the cell materials synthesized by the green and purple photosynthetic bacteria are future substrates for the chemosynthetic anaerobes which again produce the nutrients for the phototrophic bacteria. Thus these two types of bacteria growing in an O_2-free environment can exist together.

Geochemical evidence suggests that the atmosphere of the early earth (prebiotic atmosphere) was oxygen-free and consisted of H_2, N_2, CH_4, NH_3, H_2S, H_2O, CO_2, etc. The ancestral photobacterium is assumed to have existed about three thousand million years ago when the earth was already two thousand million years old. In such an environment the ancestral bacterium would have utilized light to produce energy by a cyclic electron transport system. Since the oxygen of the present day atmosphere is of biological origin (the result of the photosynthetic activity of algae and plants) the green and purple bacteria represent very ancient surviving lines of organisms which at one time were the only forms able to assimilate radiant energy in the primitive, oxygen-free atmosphere.

One of the tools the biochemist uses nowadays to trace the evolutionary relationship between various taxonomical orders of organisms is a comparison of the amino acid sequence of a particular protein which is ubiquitously distributed in all these organisms. The iron–sulphur protein, ferredoxin, is found in many fermentative bacteria and in all photosynthetic bacteria and plants. By comparing the amino acid composition and sequence of ferredoxins of various species we are in a position to propose that the photosynthetic bacteria would occupy an intermediate period in the floral evolutionary history between the ancestral fermentative anaerobic bacteria and the more recently evolved algae and plants (Fig. 7–5).

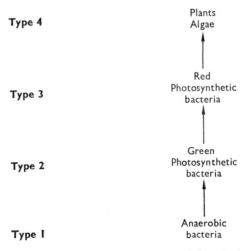

Fig. 7–5 Evolutionary development of ferredoxins.

This book is a brief synthesis of the state of knowledge of photosynthesis as we see it today. But as with many fields of biology the field is changing rapidly as research progresses with newer ideas, new techniques and better equipment. Photosynthesis is an exciting field of research since it spans interests including field ecology, physiology, biochemistry, biophysics, and even solid state physics. Research workers from many different backgrounds have a common interest in solving problems in photosynthesis utilizing very many different techniques of varying degrees of sophistication so that a nearly unique interchange of ideas and research occurs in the laboratory and at photosynthesis congresses.

We want to just briefly summarize the topics of research in the broad field of photosynthesis which are taxing research workers at present and which may continue to do so in the future.

Origin and development of chloroplasts The differentiation of proplastids in dark-grown (etiolated) plants and algae during the greening on subsequent exposure to light—the characteristic chloroplast lamellae and grana develop with their specific enzymes for photophosphorylation and CO_2 fixation.

The electron transport sequences in cyclic and non-cyclic photophosphorylation. The schemes given here are most certainly not completely correct and must be modified as more is learnt about the reaction kinetics of individual components and hitherto unknown components. The reactions involved in O_2 evolution at Photosystem II, the primary electron acceptors, and the composition of the reaction centres are only little known. The orientation of the individual components in the membrane is also of undoubted importance.

Mechanism of ATP formation The relationship of the proton pump and membrane field gradient (Mitchell chemiosmotic hypothesis) to the formation of ATP is being actively investigated using subchloroplast membrane particles, antibodies, algal mutants, inhibitors and uncouplers, coupling factors and ATPase enyzmes, electron transport control parameters in whole and broken chloroplasts, and various other techniques. The sites of ATP formation are also not clearly defined.

Transport of ions Chloroplasts 'pump' various cations, e.g. Ca^{2+}, Na^+, K^+, and anions, e.g. Cl^-, as an important part of the movement and storage of ions in the cell. We know very little of the mechanisms involved.

Carbon dioxide fixation The path of CO_2 to sugars is well documented but we know little of how a single intact chloroplast controls the passage of CO_2 into the chloroplast itself and how the products move out through the chloroplast envelope into the cytoplasm of the cell. The control of CO_2 fixation reactions may be altered artificially so as to induce plants to produce less starch and more protein to make them better food sources. Different methods of fixing CO_2 efficiently exist in some tropical grasses (C_4 plants) and their morphological and biochemical properties are being actively investigated.

Chloroplast structure Controversy still rages as to the structure of the membranes—where the chlorophyll is located, how the lipids are oriented, where and how the proteins are located in the membrane, etc. Newer techniques such as freeze-fracturing and freeze-etching and also the stereo-scan electron microscope may help to solve some of these problems. Attempts are also being made to reconstruct a chloroplast membrane starting from an artificial lipid bilayer and adding purified chloroplast proteins, pigments, etc. on to the artificial membrane.

Nitrogen fixation and nitrate reduction Both of these very important plant processes require photosynthetically produced energy sources, e.g. ATP and reduced ferredoxin, in order to operate. Much effort is now going into solving these problems.

Whole plant studies In the field of ecology and agriculture much work is being done on distribution patterns affected by photosynthetic efficiency, e.g. tropical grasses, desert plants. Environmental pollution is now serious, e.g. eutrophication of lakes, smog in city air, toxic chemicals in the sea, etc., and very often the initial effects of the pollutants are on the photosynthetic abilities of algae, plants and phytoplankton.

9.1 Relationship between chlorophyll content, starch synthesis and CO_2 fixation

(a) Use of variegated leaves from *Pelargonium* or *Coleus*; test for starch by iodine reaction or for photosynthesis by leaf-disc technique (PAGE, p. 45).

(b) Light- and dark-grown plants; test for starch by iodine reaction after methanol extraction (BOWEN, p. 147).

(c) CO_2 requirement for starch synthesis; plants grown $\pm CO_2$ and starch detected by iodine assay (PAGE, p. 44).

(d) Bicarbonate concentration in solution and rate of photosynthesis; leaf-disc technique (PAGE, p. 45).

9.2 Photosynthesis in *Elodea* and algae

(a) O_2 evolution by *Elodea*; influence of light intensity and temperature; use Audus microburette to measure amounts of O_2 given off (MACHLIS and TORREY, p. 132).

(b) CO_2 uptake by *Elodea*; light requirements; use of indicator dye (bromthymol blue) to measure uptake (BOWEN, p. 141).

(c) O_2 evolution by algae, e.g. *Chlorella*, *Scenedesmus*; measured in Warburg manometer (DUNN and ARDITTI, p. 36).

9.3 Separation of chloroplast pigments and chromatography

(a) Pigment extraction; acetone extraction and separation into petroleum ether and methanol (MACHLIS and TORREY, p. 136).

(b) Column chromatography of leaf extracts; magnesium oxide powder developed with petroleum ether and benzene (MACHLIS and TORREY, p. 138).

(c) Column chromatography of algal extracts; tricalcium phosphate as the absorbent and developed with phosphate buffer pH 6 (DUNN and ARDITTI, p. 36).

(d) Thin-layer chromatography; silica gel on glass plate developed with petroleum ether and acetone (BOWEN, p. 129).

9.4 Hill reaction in isolated chloroplasts

(a) Reduction of the dye, dichlorophenol indophenol, by chloroplasts; measured with colorimeter or spectrophotometer (MACHLIS and TORREY, p. 141; BOWEN, p. 135; DUNN and ARDITTI, p. 43).

(b) Oxygen evolution in the presence of potassium ferricyanide as the electron acceptor; measured in Warburg manometer or oxygen electrode (PACKER, p. 174).

9.5   Action spectrum of CO_2 fixation in algae

E.g. *Chlorella, Scenedesmus*; measured as uptake of radioactive ^{14}C-bicarbonate in light of different wavelengths using cellophane or plastic sheets, e.g. 'Cinemoid' from Strand Electric and Engineering Co., London, WC2 (DUNN and ARDITTI, p. 38).

9.6   ATP formation by isolated chloroplasts

Measure disappearance of inorganic phosphate from reaction due to formation of ATP in the light—use of radioactive ^{32}P not necessary (DUNN and ARDITTI, p. 43).

References

BOWEN, W. R. (1969). *Experimental Cell Biology; An Elementary Laboratory Guide*, Collier Macmillan, London.

DUNN, A. and ARDITTI, J. (1968). *Experimental Physiology: Experiments in Cellular, General and Plant Physiology*. Holt, Rinehart & Winston, London.

PACKER, L. (1967). *Experiments in Cell Physiology*. Academic Press, New York.

PAGE, R. M. (1967). *Plants as Organisms: Laboratory Studies of Plant Structure and Function*. W. H. Freeman, London and San Francisco.

MACHLIS, L. and TORREY, J. G. (1959). *Plants in Action*. W. H. Freeman, San Francisco.

see also:

KIRBY, T. W. and CLARK, H. P. (1971). *Experimental Biology*. Oxford University Press.

KROGMANN, D. W. (1971). *Molecules, Measurement, Meanings*. W. H. Freeman, San Francisco.

WITHAM, F. H., BLAYDES, D. F., and DEVLIN, R. M. (1971). *Experiments in Plant Physiology*. Van Nostrand Reinhold, New York and London.

Further Reading

Non-specialist books

FOGG, G. E. (1968). *Photosynthesis*. English Universities Press, London.
HEATH, O. V. S. (1969). *The Physiological Aspects of Photosynthesis*. Heinemann Educational Books, London.
RABINOWITCH, E. and GOVINDJEE, (1969). *Photosynthesis*. John Wiley, New York.
TRIBE, M. and WHITTAKER, P. (1972). *Chloroplasts and Mitochondria*, Studies in Biology No. 31. Edward Arnold, London.

Scientific American offprints

ARNON, D. I. (1960). *The Role of Light in Photosynthesis*.
BASSHAM, J. A. (1962). *The Path of Carbon in Photosynthesis*.
LEVINE, R. P. (1969). *The Mechanism of Photosynthesis*.
LEVINE, R. P. and GOODENOUGH, U. W. (1970). *The Genetic Activity of Mitochondria and Chloroplasts*.

More specialized books and articles

ARNON, D. I. (1971). The Light Reactions of Photosynthesis. *Proc. Natl. Acad. Sci. U.S.*, **68**, 2883.
BASSHAM, J. A. (1971). Control of Photosynthetic Carbon Metabolism. *Science*, **172**, 526.
BRANTON, D. (1969). 'Membrane Structure'. *A. Rev. Pl. Physiol.*, **20**, 209.
CLAYTON, R. K. (1970). Light and Living Matter, Vol. 1: *The Physical Part*. Vol. 2: *The Biological Part*. McGraw-Hill, New York.
GIBBS, M., ed. (1971). *Structure and Function of Chloroplasts*. Springer-Verlag, Berlin.
GREGORY, R. P. F. (1971). *Biochemistry of Photosynthesis*. Wiley-Interscience, London.
HALL, D. O. and WHATLEY, F. R. (1967). The Chloroplasts, Chapter 4 in *Enzyme Cytology* ed. ROODYN, D. B. Academic Press, London.
HALLDAL, P. (1970). *Photobiology of Microorganisms*. Wiley-Interscience, London.
HATCH, M. D. and SLACK, C. R. (1970). Photosynthetic CO_2 Fixation Pathways. *A. Rev. Pl. Physiol.*, **21**, 141.
HATCH, M. D., ed. (1971). *Photosynthesis and Photorespiration*. Wiley, New York.
HILL, R. (1965). The Biochemists' Green Mansions: the Photosynthetic Electron Transport Chain in Plants, in *Essays in Biochemistry*, Vol. 1, Academic Press, London.
KIRK, J. T. O. and TILNEY-BASSETT, R. A. G. (1967). *The Plastids*. W. H. Freeman, London.
LEHNINGER, A. L. (1970). *Biochemistry*. Worth, New York.
METZNER, H., ed. (1969). *Progress in Photosynthesis Research*, Vols. I, II and III. Int. Union Biol. Sc., Tübingen, Germany.
OLSON, J. M. (1970). The Evolution of Photosynthesis. *Science*, **168**, 438.
PARK, R. B. and SANE, P. V. (1971). Distribution of function and structure in chloroplast lamellae. *A. Rev. Pl. Physiol.*, **22**, 395.
RAVEN, J. A. (1970). Exogenous Inorganic Carbon Sources in Plant Photosynthesis. *Biol. Rev.*, **45**, 167.
RUHLAND, W., ed. (1960). *Encyc. Plant Physiology*, Vol. 5. Springer, Berlin.
VERNON, L. P. and SEELY, G. R. eds. (1966). *The Chlorophylls*. Academic Press, New York.
WALKER, D. A. (1970). Three Phases of Chloroplast Research. *Nature, Lond.*, **226**, 1204.